DIGITAL PHYSICS

DIGITAL PHYSICS

The Meaning of the Holographic Universe and Its Implications Beyond Theoretical Physics

EDIHO LOKANGA

Enquiries should be addressed to:
Buku_masapo34@yahoo.com

Printed in the United States of America

Book covers and design by:
Wordzworth Book Designers and Publishers
www.wordzworth.com

ISBN-13: 978-1979869515

DEDICATION

This book is dedicated to the seven million people killed during the civil wars in the Democratic Republic of the Congo. Innocent civilians were murdered because of the infighting (funded by multinational companies) to grab the resources of the rich Congo. Let us come together and join hands in a show of solidarity to wipe out poverty, suffering, injustice, discrimination, and humiliation that the people of Eastern Congo have gone through.

Millions of people have lost their lives, homes, and belongings for reasons they did not create; it breaks my heart. The people of the Congo say:

- No to imperialism
- No to domination
- No to discrimination
- No to injustice
- No to land grabbing

The oppressed workers and peasants of the Congo say enough is enough. The Congolese population seek peace, equality, justice, and an end to the constant exploitation of their resources.

CONTENTS

ILLUSTRATIONS

FIGURES

TABLES

ABBREVIATIONS

Abbreviation	Meaning
AdS/CFT correspondence	anti-de Sitter Space/Conformal Field Theory Correspondence
BH	Black Hole
CDM	Cold Dark Matter
CMBR	Cosmic Microwave Background Radiation
DM	Dark Matter
FT	Fourier Transform
GR	General Relativity
HP	Holographic Principle
HU	Holographic Universe
MOND	MOdified Newtonian Dynamics
QG	Quantum Gravity
QH	Quantum Hologram
QI	Quantum Information
QM	Quantum Mechanics
QT	Quantum Theory
QTG	Quantum Theory of Gravity
SM	Standard Model
ST	String Theory
ToE	Theory of Everything
WMAP	Wilkinson Microwave Anisotropy Probe
ZPE	Zero-Point Energy

ACKNOWLEDGMENTS

I would like to express my appreciation to the many people who continue to assist and help me daily. Special thanks to my wife Charlotte Lokanga and my children Engombe Wedi Lokanga, Shungu Umumbu Lokanga, Maranga Pama Lokanga (Mama Tshike), and Onadikondo Omadjela Lokanga (Papa Tshike). I extend my gratitude to my father Chief Lokanga Lopongo Onadikondo and my "mothers" Ehadi Walo, Ombawo Mundeke, and Ohandjo Tendake. My highest esteem goes to my late grandparents Chief Onadikondo Omadjela, Chief Ediho Kengete Ta Koi, and Mama Esambo Wanyakoi.

I am very grateful to Professor Rene Yamapi, Professor Laurent Cleenewerck and Euclid (Euclid University) for supporting my research in theoretical physics and theoretical computer science. Thanks also to my dear brother and friend Professor Ernest Parfait Fokoue, who continues to give moral support and practical assistance, and from time to time through our correspondence helps me to clarify ideas and provides constructive criticism.

My appreciation extends to my brothers and sisters for their interest, advice, and encouragement. Thanks to my brothers Onadikondo Omadjela, Lokanga Lopongo, Tsheko Lopala, Tolosudi Lokanga, Mukanga Lokanga, Tokombe Lokanga, Mupe Ndombasi (Ne Ndombasi Kongo), and Istadeva Emmanuel, for their constant support and encouragement. To my sisters Esambo Wanyakoi, Kongadji, Wondo Handaki, Dondja Esenge, Kabibi Lokanga, Atakete Lokanga, Omba Lokanga, Shako Lokanga, Mbutshu Lokanga, Mama Tshike Lokanga, and Okashoko Lokanga, I say a big thank you. I also offer my gratitude and thanks to all members of my extended family.

Finally, I would like to pay my highest respects to and express admiration for the work done by the following people:

- Cheikh Ahmadou Bamba, a Sufi religious leader and the founder of the large Mouride Brotherhood in Senegal.
- Professor Ngũgĩ wa Thiong'o, the most important African writer.
- P.E. Lumumba, Democratic Republic of the Congo, father of the nation

INTRODUCTION

Throughout the history of science there have been many discoveries of new phenomena, which have not been satisfactorily explained by the prevailing theories or existing physical laws. For instance, despite the successes and achievements of quantum theory (QT), some paradoxes, mysteries, countless interpretations, different schools of thought, gaps in experimental work, or incompleteness of the theory have been pointed out. Something is missing from those theoretical and experimental studies. Some arguments or explanations seem incomplete or perhaps unreliable. As a result, new hypothesizes and theories arise. They emerge out of the prevailing wisdom and are postulated to cover new knowledge or explain the phenomena, failures, or incompleteness of the old theory in a new and, perhaps better, framework.

Therefore, the way forward would be for new experiments to be designed and performed until a satisfactory outcome was obtained come out of the theoretical hypothesis and the experimental work; new theories could be accepted and lead to new physical laws—thus collaboration among theorists, engineers, and experimentalists is crucial. We have learned throughout the history of science and technology that science is an evolving process—nobody knows what the future will bring.

Physics has gone through a revolution in recent centuries, from classical physics to QT, Einstein's theory of relativity, quantum gravity (QG), etc. With the advent of Einstein's theory of relativity, we have learned that physical reality is inseparable from the spatial structure. For instance, space and time, which were previously considered as two different independent entities, are now seen to be relative—related to one another through the underlying spatial structure of space-time. Einstein's special theory of relativity taught us that space and time form a four-dimensional continuum (space-time), and that time is relative, being neither linear nor absolute. Einstein's work on relativity revealed the remarkable interconnectedness of space and time.

Since olden times, the notion of space and time has had different meanings in different world cultures. There are, in fact, different frameworks of reality in all the old traditions and cultures of the world. For instance, in the culture of the Native Americans, there is no such thing as a clock in the modern sense. Time is divided into the *Now* and *All Other Time*. Likewise, the Aboriginal people of Australia use a different framework of time which is split into two phases: the *Passing Time* and the *Great Time*. The later has a sequence but cannot be dated.

Astronomical evidence shows that by 2773 BC the Egyptians had instituted a 365-day calendar. Many centuries later, Hellenic astronomers added the missing one quarter day to the Egyptian calendar. The Romans under Julius Caesar adopted the calendar with a leap day in 46 BC. Since then, the calendar has had one modification, applied under Pope Gregory in 1582 upon the advice of astronomers. This concept of time borrowed from the Egyptians spread to the western hemisphere, where the notion of absolute parameters for recording time has become very crucial. Hence, a *clock* or a *calendar* is used.

This notion of time has spread and been adopted by the rest of the world, so that in our modern world a year is made up of 365 days, a day contains 24 hours, an hour consists of 60 minutes, and a minute is made up of seconds. There is no doubt that the use of absolute parameters has become crucial since the industrial revolution. Since the development of QT and Einstein's work on relativity, we began to witness the development of other related physical sciences such as particle physics, chemistry, biophysics, the quantum hologram (QH), information theory, thermodynamics, etc. With the rise of those subjects, we started to see the interdependence, interconnectedness, and the obvious links that bind all scientific disciplines. Moreover, as a result, we are getting a better objective description, or picture, of our universe.

It is worth pointing out that despite the progress that has been made, it is very unfortunate that we continue to depict, describe, and interpret the laws of the universe as if we were still living in a purely material-mechanical system. We continue to depict the world and the universe as a giant mechanical system, running according to Newton's laws of motion. It is important to remind ourselves that one of the most important features of the quantum model is that electrons, as well as other subatomic particles, are not objects at all. One popular interpretation of quantum mechanics (QM), a theory that attempt to explain physical behaviours at the atomic and subatomic levels (very small things), stresses the behavior of an electron that can manifest itself a wave or a particle. The electron manifests itself as a particle only when it is being observed. In contrast, when it is not being observed, the outcome of experiment suggests that it is always a wave. Also, string theory (ST), the most likely promising candidates of the theory of everything, proposes that subatomic particles are strings, so tiny that they appear to us as points informs us that all particles are just tiny loops of vibrating string.

This book looks at the deeper meaning of the word holography, its association with the universe, and its impact on physics and beyond. In a nutshell, it explores the old conundrum of whether or not we are living in a holographic universe (HU) and its possible consequences. An attempt is made here to convey to the wider audience insights emerging from the forefront of theoretical physics research, which are accessible to a broad spectrum of readers, especially those with no training in physics or mathematics. This topic has wider implications—it highlights the connections among QT, non-locality, the QH, quantum holography, QG, and Bell's theorem.

There has been a fascination with holography since the early 1970s. This subject is central to all areas of science and has for many years been the source of much confusion. Although holography entered science with the work of the Hungarian scientist Denis Gabor, it is only in the last 40 years that it has developed into a subject in its own right. David Bohm, more than anyone else, is responsible for this advance. The concept of the HU was initially developed, extended, and promoted by two of its most ardent defenders, Bohm and the neurophysiologist Karl Pribham. Their holographic model received much attention and recognition after dramatic experimental support for it emerged in 1982. A research team, led by physicist Alain Aspect in Paris, demonstrated that the web of subatomic particles that compose our physical universe might possess a holographic property.

Bohm suggested that particles can communicate with each other because the separation between them is an illusion. He argued that the separateness is a projection from a higher level of reality that he calls *Implicate Order*, where everything relates to everything else. The particle behaves as if it knows the other particle because it has the other particle's information within itself. One the other hand, Pribham, when attempting to find which specific locations in the brain were assigned to store our memories, also saw holography in action. His experiments revealed that there is no localization of memory. Pribham claims that the pattern formed by the interference of electrical signals from each nerve cell in the brain is where, in fact, memories are 'stored'.

In the theory of black hole (BH) thermodynamics, the concept of a HU may also have emerged through theoretical models in physics, which attempt to explain the action of the four fundamental known forces in nature. Subsequent work in the physics of BHs has also contributed, starting with the work of Stephen Hawking and Jacob Bekenstein and culminating with the publication of "Dimensional Reduction in Quantum Gravity" by Gerard 't Hooft[1] and Leonard Susskind[2]. "The World as a Hologram," meant that the holographic nature of all physical systems became apparent, for it explains how a three-dimensional physical system can be described by a theory based on a system's two-dimensional surface area.

The deep connection between physics and information was highlighted in the early 1970s when the link started to become apparent. The late physicist John Wheeler challenged his Ph.D. student Jacob Bekenstein on what would happen if one poured a hot cup of tea through the event horizon of a BH. According to the second law of thermodynamics, information could

1 Hooft, G. 1993. "Dimensional Reduction in Quantum Gravity." Accessed October 11, 2012. http://arXiv:gr-qc/9310026.

2 Susskind, L. 1995. "The World as a Hologram." *Journal of Mathematical Physics*. 36:6377–6396.

not be destroyed or lost. So, the information had to be stored somehow, but how? This led to what is known today as the BH information paradox. A possible outcome to the paradox was to consider first that the information was never lost in the BH but was instead stored on the event horizon. The work of Bekenstein led him to develop what is known today as the Bekenstein bound. Subsequent work on BH thermodynamics opened up a new chapter and resulted in the concept and development of Bekenstein-Hawking radiation. The thermodynamics of BHs is a completely new way of looking at fundamental physics, where information plays a crucial role and is paramount to any theory.

With the advent of QT, we now have a deeper understanding of the atom. We are now armed with the knowledge that matter is made up of atoms, which are in turn made up of electrons, protons, neutrons, and quarks, etc. All of these small entities are discrete particles that can be counted. For instance, an atom of copper has 29 positive charges on its nucleus and 29 electrons surrounding the nucleus. A hydrogen atom has one electron. In the same physics, other essential elements such as space and time are continuous. The later simply means that time and space cannot be divided into a number or be counted; one can only measure the extent of time that has elapsed, but not count it.

This dilemma might be resolved very soon as current advances in QG suggest that time and space may ultimately be found to be discrete and not continuous after all. Most of the theories of gravity are based on entropy and holographic arguments—there is an overwhelming feeling that gravity might not be a fundamental force but rather an emergent phenomenon. This opens the door for physics to be reformulated in a new way, so that information is essential for physics and the laws which govern it could perhaps be derived from that foundation. Recent advances in theoretical physics suggest that physics is fundamentally based on information. If this is the case, we would like to indicate that the universe is digital or discrete, or to put it simply, the universe computes. Furthermore, the holographic principle (HP) also suggests that the concept of information is as fundamental as matter or energy.

The universe is digital, or it computes. It is a giant computer processing reality. Each time it processes it has access to other information and more details to build upon and it is continually evolving. Digital physics casts a wide net—it is a broad, new, deep, all-encompassing subject that draws on many of the branches of physics, mathematics, and computer science. Viewing the information as fundamental has led some physicists to propose the placing of information at the core of physics. With the holographic model of the universe, we have come to realize that every individual point in the universe contains the whole universe itself. As a result, every point in space is connected to another point further away in space, and every subatomic particle comprises a web of interconnections by which it becomes intertwined with other parts of the universe. This view opens possible avenues of understanding the universe in a way we never suspected before. We learn to understand the interconnectedness of all things; thus, we gain new insight concerning the nature of reality about QT and beyond.

Finally, I hope that this book will also be of interest to those readers who do have a scientific background. For the avid reader of popular science books, I have tried to explain difficult words; a glossary of scientific terms is included at the end of the book. Furthermore, I have avoided technical language and equations—the notes section can be skipped with minimal impact on the book's logical flow. The glossary of scientific terms, for an easy and accessible reminder of ideas found throughout the book, should be useful for those new to this subject.

Ediho Kengete Ta Koi Lokanga,
Tipton,
West Midlands,
United Kingdom,
January 2018.

PART ONE

A Holographic Universe

A Holographic Universe?

One thus sees that a new kind of theory is needed which drops these basic commitments and at most recovers some essential features of the older theories as abstract forms derived from a deeper reality in which what prevails in unbroken wholeness.

—DAVID BOHM

HOLOGRAPHY IN A NUTSHELL

This chapter is about holography, and how the concept of holography is connected to the information paradox. The idea of holography is well supported and developed in certain models, particularly in brain research, cosmology, QG, QH, and ST in the context of anti-de Sitter space/conformal field theory (AdS/CFT) correspondence[3]. As such, a more general formulation is lacking, and to some degree, the ultimate role of the HP in fundamental physics remains to be identified. What do we mean by a HU? What is the meaning

3 In 1997, Juan Maldacena made a remarkable discovery that showed a connection (based on the HP) between the anti-de Sitter space /conformal field theory correspondence (AdS/CFT correspondence), a five-dimensional anti-de Sitter space and its boundary, and a four-dimensional space that possesses a conformal field theory (quantum gauge theory). "String theory in a particular curved ten-dimensional space-time is equivalent to a gauge theory in flat four-dimensional space-time." The five-sphere is a five-dimensional generalization of the two spheres—it is a symmetrical space with a constant positive curvature. On the other hand, anti-de Sitter is a space-time with constant negative curvature. As a matter of fact, the full space-time has ten dimensions, five lie along the five-sphere and the other five spheres are along five-dimensional anti-de Sitter space. The conjecture demonstrates that the gauge theory is the hologram of the ST.

of HP[4] in the context of QG theory and QM? What impact does it have to our understanding of life? This is a pertinent issue because the HP states that there is a precise and strong limit on the information content of space-time regions. Moreover, the number of quantum states in a spatial region is bounded from above by the surface of the region measured in units of four-fold Planck areas. As a result, this entropy bound places a strong constraint on any theory of our universe. Moreover, if this principle is proven to be true, then due to the overwhelming evidence, ST and field theory cannot be considered as ultimate theories of nature.

The HP is not yet generally accepted by most physicists, even though it has been demonstrated theoretically, and there is mounting evidence and support for it[5]. To properly understand the principle, we would look at the meaning of entropy in a wider context, and show that entropy or information is crucial to our understanding of the HP. In the following lines, we attempt to explain what this principle is, and how it could unlock the mysteries and paradoxes of physics.

We also show that most of the fundamental problems that have been identified over the last 70 years or so might be understood from the perspective of a HU. Moreover, in doing so, we stress the fact that the HP might be one of the missing pieces of the puzzle for the formulation of a theory of everything (ToE). Let us illustrate the concept of a HP through a hologram. A hologram is a two-dimensional plate or piece of film that creates a three-dimensional image when it is illuminated. All the information needed to create this image is contained within the plate. This principle has been extended to any system occupying some dimensional regions. For instance, a physical theory

---

4 The HP states that there is limit on the information content of space-time regions. The number of quantum states in a spatial region is bounded from above by the surface of the region measured in the unit of four-fold Planck areas.

5 A team of Japanese physicists has provided what might be the clearest evidence yet that our universe could be a giant hologram. This idea is like that of holograms, where a three-dimensional image is encoded in a two-dimensional surface. According to the HP, gravity comes from thin vibrating strings. The strings are holograms of events that take place in a flatter cosmos. The HP suggests that there is a two-dimensional surface that contains all the information needed to describe a three-dimensional object—in this case, our universe.

defined by the two-dimensional boundary of the region can fully describe the physics of the three-dimensional region itself. This means that what we see as a three-dimensional system may just be an illusion, a projection from a two-dimensional system operating within another set of physical laws on the boundary of that region.

From the viewpoint of the hologram, it is hard to distinguish between a three-dimensional region and a holographic projection from a two-dimensional surface. For instance, in the case of our universe, if we live in a four-dimensional universe, i.e., three dimensions of space plus one of time—this could be a holographic projection built into a three-dimensional boundary of our universe, with two spatial dimensions and time. Remarkable theoretical advances were made in 1997 when Juan Maldacena made an astonishing discovery. He showed a connection between the anti-de Sitter space/conformal field theory correspondence (AdS/CFT correspondence), a five-dimensional anti-de Sitter space (ST) and its boundary, and a four-dimensional space that possesses a conformal field theory (quantum gauge theory).

> *"String theory in a particular curved ten-dimensional space-time is equivalent to a gauge theory in flat four-dimensional space-time."*

Despite many years of research, the fundamental mechanism of the AdS/CFT remains a mystery to many physicists. Which region of AdS is responsible for and behind the mechanism of information in the dual CFT?[6] In a contribution to this long-standing problem, I would like to investigate, explore, formulate,

---

[6] The Maldacena proposal provided physicists with a dual mathematical tool and temporarily solved the apparent inconsistencies between quantum theory and Einstein's theory of gravity. It provides a duality to move back and forth between the two theories. Although his ideas have been developed and expanded, they have perhaps been taken for granted. A Japanese physicist, Yoshifumi Hyakutate, and his colleagues provided the first compelling evidence that the Maldacena conjecture might be true. This is the first time that the validity of the model proposed years ago has been mathematically tested. Commenting on this, Maldacena says "The whole sequence of papers is very nice because it tests the dual [nature of the universe] in regimes where there are no analytical tests." Another researcher well known for his work on gravity, Leonard Susskind, added that "They have numerically confirmed, perhaps for the first

and try to understand the HP in a wider context, by examining various aspects, views, and approaches in different areas of science. Furthermore, I would like to explore the holographic description of the universe, and then review its current status in various areas of science. I will attempt to show that *information*, not field theory, is behind every action in physics. I argue that the physical processes of the universe should be understood in terms of *information exchange* rather than the interaction of fields.

What is information? It is not self-explanatory, it only exists as a potential, and it requires a medium to manifest itself. Hence, it is a difficult to define what information is. We all know that information can be transmitted or communicated instantly; one can get direct information about the properties of a physical object such as a star in the sky or a tree. Even though a star is located billions of light years away from us, by observing it we perceive the information generated by the object (such as its colors, shape, size, composition, etc.). Around us and throughout the universe, we perceive information generated by multiple objects; the universe can be considered as a potential pool of information comparable to a holographic informational universe—the universe is thus made up of information.

In information theory, any variable which can assume the values 0 or 1, yes or no, is called a binary digit or simply a bit. We can use binary digits such as 0 or 1 to transmit a set of instructions. Instructions to make a binary choice are simply given by transmitting yes, for instance, to suggest hot or 0 to suggest cold. Briefly, some bits of information can be encoded in a physical system; this is true when, for instance, a set of instructions in the form of n binary choices need to be transmitted to identify the state of a physical system. Information is always encoded in a physical system—the erasure of information leads to an increase of the entropy of the universe; this is known as Landauer's principle. Physicists have come to the view that any information can be encoded, then processed, and finally transmitted by physical means.

---

time, something we were fairly sure had to be true, was still a conjecture- namely that the thermodynamics of certain black holes can be reproduced from a lower-dimensional universe."

There is not a single or universally accepted definition of information. Sunil Thakur[7] defines information as "any form of radiation that communicates one or more properties of a physical entity to another physical entity." On the other hand, Stonier[8] argues that" information is the cosmical organizational principle with a status equal to matter and energy." Di Biase and Rocha[9] have gone a step further and view information as "an intrinsic, irreducible, and non-local, property of the universe, capable of generating order, self-organization, and complexity." However, others, like Chalmers[10], see information and consciousness as one entity. He shares the view that consciousness is "an irreducible aspect of the universe, like space and time and mass." There are in fact many definitions of information, depending on which school of thought one is coming from. In the light of the above remarks, we think that a complete definition must consider the perspective of information theory, consciousness, and QH, particularly the non-local effect, mainly the instantaneous transmission of information that does not involve bits or qubits.

The development of science out of its African (Egyptian and medieval) origins marks the beginning of the separation between the physical and non-physical realms. This separation represents the triumph of rationality and the empirical approach to understanding the nature of the universe. The history of science tells us that during the development of science in Egypt, as well as during Plato's and Aristotle's time, the physical and non-physical realms existed side by side. However, during the eras of Descartes and later Sir Isaac Newton, western science rejected the notion of the non-physical hypothesis.

The triumphs of modern physics took precedence over the concept of a non-physical hypothesis, to preserve the idea of physical contact between particles and

---

7 Thakur, S. 2010. "Quantum Entanglement and Holographic. "Accessed April 12, 2010. http://www.norlabs.org/

8 Stonier, T. 1990. *Information and the Internal Structure of the Universe.* Springer-Verlag, New Addison-Wesley: Reading, MA.

9 Di Biase, F. and Rocha , M. 1999. "Information Self-Organization and Consciousness. "*World Futures- the Journal of General Evolution* 53:309–327, UNESCO, The Gordon and Breach Group, U.K.

10 Chalmers, D. J. 1995a. "Facing up to the Problem of Consciousness." *Journal of Consciousness Studies* 2(3):200–19.

to explain phenomena such as gravitation, electromagnetism interactions, weak and strong nuclear forces, etc. These ideas of modern physics include quark theory, the field theories, the STs, QT, the Standard Model (SM) of matter and energy interactions, as well as the success of the big bang cosmological models that explain the origin and development of the universe. However, it is worth pointing that these successes have also lead to the increasing realization that several fundamental questions remain unanswered from a scientific point of view.

Today, many scientists who have been struggling at the cutting edges of their fields are calling for a more inclusive approach to science, and have come up with concepts that resonate with those of earlier times. For instance, QT stresses the fundamental importance between observers and the observed, and that matter is not only made up of atoms but exists as the relationship of energy, matter, information, and vibrations. Likewise, in the medical field many leading physicians have argued that the healing process involves the whole person, i.e., body, mind, and spirit.

These ideas of wholeness have been taught for generations in the cosmology of indigenous peoples of Africa, Asia, America, and even Europe; there is no separation between matter and spirit, or between the individual and society, or us and nature. More and more people are calling for the need to recognize this interconnectedness. Thus, in the spirit of wholeness, science must encompass matter and spirit, and not deal with matter in isolation. This is similar to how we have fragmented space from time in cosmology. Today the so-called ToE is beginning to open itself up to new possibilities. From within this openness, a new way of accommodating new ideas coming from the science of information may be possible.

Information has strongly re-emerged in the non-local[11] form in quantum physics. Arguments that validate the theory and experiments of the

---

[11] Entanglement is a phenomenon that has been demonstrated repeatedly through experiment. Einstein called it "spooky action at distance". The mechanisms behind this phenomenon have not yet fully been explained by any theory. Nonetheless, several schools of thought and competing theories are trying to give a plausible explanation and a satisfactory answer to what some people prefer to call quantum non-locality. Entanglement is a very strange type of thing, obscure and mysterious; like objects

non-local QH[12] are well supported and have given information and non-local information a critical status in physics. Information (patterns of energy) are the basis of cognition and the mechanisms by which beings and other creatures perceive reality. The non-local effect is a useful property of the universe, and many phenomena associated with our daily experience relate to the phenomena of non-locality. Science has not widely accepted the explanatory physical mechanisms behind non-locality[13].

How do we humans perceive information about the physical object which is not available through local, classical space-time, or neural mechanisms? There is an overwhelming abundance of evidence that we can perceive information about objects which are not available locally. We need to explore the science behind and around holograms and examine what QHs say about our universe[14]. The QH may explain the instantaneous communication that

---

being separated in far lands while still being in communication with each other, without using any obvious means of communication. Some would call this telepathy. During entanglement, there is no energy transmission between particles or objects, only information exchange. Entanglement simply means that we are all connected and that there is no spatial separation between objects or one particle and another.

12 Edgar Mitchell argues that the concept of the QH is based on quantum emissions from all physical objects: you, me, the camera. Any physical object of macroscopic size, molecular and above, emits and absorbs quanta of energy. The quanta emitted from every object we've discovered carry information about the physical. The QH is this informational structure about a physical object and it is non-local, which means it is not space-time restricted.

13 Perhaps non-locality and the non-local QH provide the mechanism which offers a plausible solution to the host of enigmatic observations which experiments associate with consciousness. We learn from QH that each physical object is associated with the non-local wave form well known in quantum theory (wave-particle duality). All objects, all physical matter, in our universe are associated with non-local wave forms which are functioning at all levels. The possible existence of non-local QH make it possible to explain how the universe has made use of non-local information from the beginning, to self-organize and permit complex life to evolve, including space-time.

14 According to the QH theory, we live in a three-dimensional world embedded in a HU of infinite dimensions—we beings are perceivers of the quantum universe and as well as receivers of the information coming from the QH. We decode frequency information received from the quantum holography into our three-dimensional world. As such, the QH takes precedence over our three-dimensional world. Everything forms in the QH before anything manifests itself in our world—the QH is the blueprint that determines and builds our three-dimensional world.

exists between all entities in the universe and possibly shed light on how quantum entanglement is possible.

In 1982 Aspect and colleagues performed an experiment that changed the way we think about entanglement and holography. They discovered that, under certain circumstances, subatomic particles could instantaneously communicate with each other regardless of the distance separating them—each particle seems to know what the other is doing. Following Aspect's experiment, many other similar tests have been conducted. For instance, in 2008 Nicolas Brunner and colleagues at the University of Geneva, Switzerland, conducted a Bell-type experiment with human detectors. The authors stressed that such an experiment would probably not lead to a better understanding of quantum non-locality itself, but it would be fascinating! The outcome of this experiment suggests that the human eye can instantaneously detect action at a distance. In a nutshell, the experiment indicates that it is possible that two eyes, i.e., two people, get entangled and that no physical link is required between the observer and the observed.

Although the experiments mentioned do not conclusively mean that faster-than-light communication between beings or particles is possible, they do hint that we can communicate with one another regardless of the distance between us. These findings show that our notion of matter and physical objects may not be correct and that objective reality does not exist. In the light of various experiments in a QH, it is worth mentioning some other exciting work that might allow us to understand holography. For instance, Ayman F. Abouraddy and colleagues made use of quantum entanglement to extract holographic information about a remote three-dimensional object in a confined space which lights enters, but from which it cannot escape. The underlying entanglement permits the measurement to yield coherent holographic information about the remote object. A light beam attains the remarkable possibility of quantum holography that itself does not illuminate the object, but is entangled with the beam that does, and is detected with full spatial resolution.

In the same spirit, Paolo Manzelli proposed a simple speculative model to make use of quantum entanglement for extracting holographic information in the occipital brain for remote three-dimensional perception. Manzelli's work suggests that information energy is the causal link that permits the coexistence of non-local effects. It follows that the idea of the brain as a Turing machine, incapable of supporting quantum processes, can no longer be accepted. The works of Roger Penrose and Karl Pribham have renewed interest in the need to uncover the quantum processes taking place in the brain. Theoretical work and experimental evidence has shown that the brain process information holographically.

In QG, the concept of the HU continues to gain ground. In January 2010, an idea was put forward by Erik Verlinde of the University of Amsterdam, suggesting that gravity (long considered as a fundamental force) is, in fact, an emergent phenomenon, and arises simply from the complex interaction of simpler things[15]. Verlinde argues that gravity is simply a manifestation of entropy in the universe. His thesis is based on the second law of thermodynamics: entropy always increases over time. Accordingly, the differences in entropy between different parts of the universe generate a force that redistributes matter in a way that maximizes entropy. This is, according to Verlinde, the force that makes gravity, suggesting that gravity is essentially a phenomenon of information.

Verlinde's work naturally raises some fundamental questions. It led me to ask: Is information the causal force of gravity? Is the universe built of information? What is the fundamental role that information intends to play in the unification of QM and relativity? A short review of space, time, and cosmology, in addition to the BH theory, should help understand the idea behind Verlinde's reasoning.

---

15 Verlinde's ideas have been independently supported by the work of Jae-Weon Lee and his colleagues at Jungwon University in South Korea. They have used a different approach, mainly QI, to derive a theory of gravity. The outstanding questions of what happens to information when it enters a BH may lead to a good understanding of gravity, through Landauer's work. According to Landauer's principle, the erasure of information leads to an increase in the entropy of the universe. Jae-Won and his colleagues think that the erasure process must occur at the BH horizon, and it is possible that space-time organizes itself through a process of self-organization: space-time maximizes entropy at these horizons, i.e., it generates a gravity-like force. This work relates gravity to QI for the first time.

- A BH gravitational force is so strong that nothing, not even light, can escape it. The structure of a BH is:
- A point at the center is called the singularity. A larger spherical area around it has a boundary called the event horizon. The event horizon is also known as the point of no return; at this stage, nothing can escape it. A relationship exists between a BH's mass and the surface area of its event horizon. We also know that the area of the event horizon increases proportionally with the amount of energy or mass that the BH consumes.

Does a BH violate the second law of thermodynamics? At first, it did seem so because the entropy of any object that fell into a BH was thought to disappear. However, a remarkable discovery by Bekenstein in the 1970s taught us that this is not the case. In fact, Bekenstein drew a parallel with the concept of entropy and remarked that during various processes that take place in BHs, such as the merging of two BHs, the total area of the event horizon never decreases. From this discovery, Bekenstein proposed that the entropy of a BH is proportional to the area of its event horizon and that a stationary BH behaves like a system in thermodynamic equilibrium. The conclusion of Bekenstein's work and the BH thermodynamics description is supported by semi-classical computation done by Hawking, which showed that a BH does indeed radiate and that this radiation is thermal.

It follows that the entropy of any object that falls into a BH is absorbed by the BH itself, and this is reflected in the increase of the surface area of the event horizon. Moreover, the increase in entropy of the BH must be equal to or greater than the entropy lost by the falling object. As shown, entropy and information are conceptually equivalent. The total quantity of information measured in bits is related to the total number of degrees of freedom of matter-energy. Regarding this, Bekenstein argues that:

> *"Thermodynamic entropy and Shannon entropy are conceptually equivalent: the number of arrangements that are counted by Boltzmann entropy reflects the amount of Shannon information one would need to implement any particular arrangement...of matter and energy."*

This statement teaches us that the difference between the thermodynamic entropy of physics and the Shannon entropy of information is in the units of measurement; the former is expressed in units of energy divided by temperature whereas the latter is measured in "bits" (binary digits) of information. So, the difference is merely a matter of convention. If the thermodynamic entropy is equivalent to information, and is measured in bits, the total quantity of bits is related to the number of degrees of freedom of matter-energy. The number of degrees of freedom of a particle is the product of all the degrees of freedom of its sub-particles. Were a particle to have infinite subdivisions into lower-level particles, then the number of degrees of freedom of the original particle would be infinite, thus violating the maximal limit of entropy density. The HP implies that the subdivision of particles cannot be continual; it must stop at some level and that the fundamental particle is a bit of information.

The HP suggests that information of the constituents of the BH is encoded in the radiation that emanates from its surface; as a result, no information is ever lost. The encoding of three-dimensional information on a two-dimensional surface demonstrates that holography is possibly one of the most powerful tools with which to understand the precise formulations of the universe. The work of both Hawking and Bekenstein in the 1970s suggests that the information stored in a BH is proportional to its surface area rather than to its volume. The theory of BH entropy suggests that the information content of any region of space is defined by its surface area, not by its volume, as previously thought. This leads to the conclusion that the volume itself is an illusion, and that in reality, the universe is a giant hologram.

Further support for the HU hypothesis came from the discovery of the Cosmic Microwave Background Radiation (CMBR) in 1965. Two scientists, A. Penzias and R. Wilson, were working at Bell Telephone Company on the Telstar/Echo satellite project at the time. They were investigating untapped radio noise found at wavelengths between a few millimeters and a few centimeters. The outcome of their investigation showed that once all known sources and origins of noise had been accounted for, there was a residual signal emanating from all directions. They obviously thought that this noise was coming from the planet Earth, the solar system, or possibly other galaxy.

These possibilities were eliminated one by one. Further research led to the conclusion that this noise was of cosmic origin, hence the name Cosmic Microwave Background Radiation. CMBR is very useful for satellites—it allows us to receive information about the entire universe at one specific location/place.

This discovery only became possible due to the invention of new detectors working at high frequencies of up to 30,000 MHz. The CMBR is similar to a *database of information* from which we can retrieve information about the entire universe in a single place. It gives us direct information about any given point of the universe. We learn that what happens at one end of the universe has a repercussion in another part of the universe: everything in the universe is interconnected. The discovery of CMBR suggests that the universe functions as a coherent unit—everything in space is directly connected to all other parts. It is possible to argue that without the wholeness it would be impossible to have instantaneous information about every part of the universe simultaneously at one location.

As such, the CMBR supports the hypothesis that the universe is a hologram; information about all parts of the universe exist in the rest of the universe simultaneously. Everything in the universe is intrinsically connected to everything else. Thus, the existence of the CMBR provides perhaps a plausible explanation for the existence of quantum entanglement, as well as for the HU. The entire universe is interconnected at the subatomic level and indeed at all levels of existence—that is why information is communicated instantaneously.

To sum up, I would like to say that Verlinde's insight about gravity being a product of entropy might finally open new avenues for the physics of information and computation. The work of Hawking, Bekenstein, and others on BH entropy has taught us that BHs radiate and have entropy, although the nature of this entropy remains a mystery. This peculiar property has led to the HP, stating that the number of degrees of freedom in any region of our universe grows only as the area of its boundary. One of the successes and detailed mathematical models of the HP is given by the AdS/CFT conjecture/correspondence, which is an explicit realization of holography.

Recent observational and theoretical advances related to the global cosmic structure may lead to a new understanding of our universe. In cosmology, observational evidence from a wide range of redshift surveys suggests that the fractal dimension of large-scale galaxy clustering is roughly $D_F \sim 2$. According to J. R Mureika[16]:

> *"If this result is correct, this statistic is of interest for two reasons: fractal scaling is an implicit representation of information content, and also the value itself is a geometric signature of area. It is proposed that a fractal distribution of galaxies may thus be interpreted as a signature of holography (fractal holography), providing more support for current theories of holographic cosmologies. In the near future, if the alleged fractal dimension turns out to be a whole number (integer) $D_F = 2$. It will be a confirmation of the presence of the holographic structure in our universe."*
>
> —MUREIKA 2007, 1

The holographic informational universe hypothesis indicates that a major shift in the scientific worldview is taking place. In the next chapter, we briefly review the physical basis of the HU, and look at the work of some of those who have contributed heavily to this subject. Based on these earlier works, I postulate and examine the evidence for the holographic informational universe.

---

[16] Mureika, J. R. 2007. "Fractal Holography: a Geometric Re-Interpretation of Cosmological Large-Scale Structure." Accessed April 10, 2010. http:// arXiv: gr-qc/0609001v2

PART TWO

The Holographic Universe Perspectives

The Holographic Universe Perspectives

Considered together, Bohm and Pribham's theories provide a profound new way of looking at the world: Our brains mathematically construct objective reality by interpreting frequencies that are ultimately projections from another dimension, a deeper order of existence that is beyond both space and time: The brain is a hologram enfolded in a holographic universe."

—MICHAEL TALBOT

THE CASE FOR A HOLOGRAPHIC UNIVERSE

The SM of particle physics has made remarkable progress over the last 50 years or so, and its successes and many amazing results are well documented. Its fundamental point is that matter particles, as well as the four basic forces between them, electromagnetic, weak, strong, and gravitational, should be understood in terms of quantum field theories. The successes of the model abound. For instance, experiments have confirmed the predictions of the SM of particle physics, and this is well illustrated in the electroweak sector as well as in quantum chromodynamics. However, despite its successes, the SM has several shortcomings which suggest that it cannot be considered as a complete description of nature. However, some of its successes should be embedded in a larger and more comprehensive framework, such as the one proposed in this book and those echoed by several researchers in digital physics, the physics of information, and computation.

One of the most significant shortcomings to date is gravity, along with the cosmological constant problem, i.e., the small but non-zero measured value of the cosmological constant. There are also some recent findings that may cast doubt on the validity of the SM. For instance, the discovery by astronomers in 1998 that the universe is expanding at an ever-faster clip (meaning that our universe was not behaving as it ought to be). It was a wake-up call to most physicists. According to the cosmological model, the big bang hurled our universe into existence; as a result, the overwhelming view was that its expansion (must) have been gradually slowing down. The result of the investigations revealed that its expansion was accelerating and we seem to be missing some 96% of its matter. This was a wake-up call to the physics community—it led some cosmologists to argue that there is an invisible and undetectable force, called dark energy, and an invisible material named dark matter (DM).

However, it the investigation into the possible existence of dark energy-matter has so far been inconclusive. One researcher, Tom Shanks of the University of Durham, UK, who has been examining the data from one of the telescopes used to establish the existence of these dark materials—the Wilkinson Microwave Anisotropy Probe (WMAP)—since 2007, has revealed astonishing results. WMAP detects the radiation afterglow of the big bang, and creates a temperature map of the universe as it appeared 380,000 years after its birth. Recently, Shanks has examined these maps and re-calibrated the data using distant microwave-emitting galaxies that appear in the WMAP data itself. His work revealed astonishing details, and suggests that the map of the background radiation contained far fewer fluctuations than previously thought and they were smooth, contradicting earlier studies.

It is the size of these fluctuations that is key in calculating the existence of DM and dark energy. Shanks' findings raise doubt about the existence of dark energy-matter. These findings may have far wider implications than anyone imagined and suggest that the entire mathematical framework currently used to explain the workings of the universe—the SM—may have to be reviewed. Before the hypothesis of DM-energy took precedence, the overwhelming view was that the universe was permeated by an all-pervading ether—an invisible fluid-like medium that light waves traveled through. In 2008 a Chinese physicist named Hong Sheng Zhao resurrected the old idea, and called it the ether dark fluid. Zhao argues that neither DM nor dark energy really exists, but that they are simply an undiscovered facet of gravity that operates at very large scales. Accordingly, the fabric of space acts like a fluid. This fluid can coagulate, compress, or expand; where matter is present, the fluid slows down and coagulates, amplifying gravity and giving the illusion of the presence of a mysterious DM-like material.

The aforementioned HP could provide answers to these questions. The proposal of the principle was greeted with much skepticism until 1997, when realizations of the idea were found in ST. The HP, considered a necessary feature of any theory of gravity, might play a crucial role in the solution of the cosmological constant problem. It may also play a role in understanding the structural aspects and workings of the universe. The notion of holography is linked to entropy, so it is important to understand thermodynamic entropy. Thermodynamics is the science of heat flow, production, and conversion of heat energy into work. It arose out of the study of heat engines, cannons, and steams engines during the eighteenth and nineteenth centuries; these were partially responsible for the industrial revolution.

The need for efficient heat engines during the eighteenth and nineteenth centuries led researchers to the realization that only a portion of the total heat pumped into an engine was useful. The rest was lost and escaped into the surrounding environment during work. Heat engines also lost efficiency over time. They quickly realized that the drop in efficiency was due to two factors: the consumption of useful heat and its conversion into waste heat. Taking into account these two factors, they concluded that a disordered system was an engine that cannot do work. As the engines became more disordered they lost efficiency. This disorder was quantified and given the name entropy.

Ironically, it is from these earlier works that the three laws of thermodynamics were derived. The second law of thermodynamics is particularly relevant to this discussion. It stipulates that the entropy of a system must either remain the same or increase, but cannot ever decrease, while the system performs work. In thermodynamics, there are three types of systems—isolated, closed, and open.

- An isolated system is a system where neither matter nor energy can enter or leave;
- A closed system is one where only energy can enter or leave;
- An open system is one where both matter and energy may enter or leave freely.

The second law is most relevant to an isolated system only, because in the case of open and closed systems, both can reduce their entropy while doing work by pumping waste energy and matter into the surrounding environment. By doing so, the entropy of the environment increases; as a result, the increase is greater than the decrease within the system. The second law of thermodynamics means that the sum of the change in entropy for both the system and its surroundings must be equal to or greater than zero.

From a thermodynamics viewpoint, the concept of entropy is a measure of the degree to which a system can do work.

In light of statistical mechanics, the idea of entropy was pioneered by Ludwig Boltzmann. He developed a new idea of entropy based on the study of atoms. He realized that a group of atoms could arrange themselves into a variety of configurations which he called states. By disorder, he referred to the fact that it would be difficult to describe the behavior of the atoms accurately, as they could achieve a wide variety of states. As such, the number of states that atoms could achieve would lead to more ordered arrangements. This state of matter led him to define entropy as being proportional to the log of the number of states. The Boltzmann entropy S, is given by the relation: $S=klogw$, where k is a constant and w is the number of states.

It follows that the statistical mechanical concept of entropy is a measure of the degree of complexity of a system.

In 1949, when Claude E. Shannon was trying to find a way to measure the amount of information in a message, he derived a formula that looked like the Boltzmann formula. He labeled the calculated information content as the entropy of a message. By doing so, he calculated the digital information quantity for any message using any series of symbols. His work showed that the more complex the message, the higher its entropy; in other words, the more complex a message was, the more information it contained. The outcome meant simply that a highly disordered message, with large entropy, contained more information than a simple, highly ordered message with low entropy.

From the information theory viewpoint, the information concept of entropy is a measure of the quantity of information a system possesses.

The ideas of entropy from thermodynamics, statistical mechanics, and information theory, are conceptually equivalent and measure the same thing. The realization is that whether one calculates the entropy of a system thermodynamically, statistically (statistical mechanics), or informationally (Shannon), the value of that entropy should be the same.

History and the Origin of the Holographic Universe

The concept of entropy is not only tied up in thermodynamics and information, but is also strongly linked to the idea of BHs and holography. Holography is changing our view of the universe and of space-time. Over the last forty years or so, almost all areas of science have benefited from the input of the holographic concept—holography changes our understanding of BHs. Its applications are found in various areas of science from QG, ST, quantum information (QI), medicine, etc. Studies of BHs occupy a prominent position in theoretical physics as their properties are central to understanding QG. Over the last 30 years or so, the fundamental questions in QG have been:

- Why does a BH have an entropy proportional to its horizon area?
- Is there any information loss due to BHs?
- How do we resolve the issues about space-time singularities?

A new idea, the HP, has emerged which could provide answers to these questions. The theories of a holographically based universe had been around for some time. They have been championed by the likes of Bohm and Pribham, a respected neurophysiologist from Stanford University. The Alain Aspect[17] experiment demonstrated that the web of subatomic particles that make up our physical universe possesses what seems to be a holographic property.

What is holography? Moreover, how does it impact on our understanding of life and the universe? The word holography is taken from optics. Its meanings in optics, QG, ST, neurophysiology, and QM, are similar in spirit but not identical. However, by combining this concept from all these different fields, a common attribute that links everything in the universe emerges naturally. That common attribute is information. Let us look at its meaning in optics and in the context of QG and ST.

A hologram is a three-dimensional image formed by the interference of light beams from a coherent light source. Besides light beams (one form of electromagnetic radiation), a hologram can also be set up using sound waves or physical waves such as seismic waves, ocean waves, etc. Experiments in holography have shown that holographic systems can generate three-dimensional virtual images. Figure 2.1 shows a laser beam falling upon an object and reflecting onto a photographic plate. A second laser beam falls on the photographic plate generating a mix of waves from the two beams. When the two waves overlap interference results. Also, once the wave interference pattern is reflected it generates a three-dimensional image of the object in space. The wave interference pattern from both beams stored in the film holds all the information (form and volume) about the object. The pattern information stored in the photographic plate is called a hologram.

---

17 Aspect, A., Dalibard, J. and Roger, G. 1982. "Experimental Test of Bell's Inequalities using Time Varying Analyzers." *Physical. Review Letters.* 49:1804. Accessed 1st May 2013 http://www.phys.ens.fr/~dalibard/ publications/Bell_test_1980.pdf

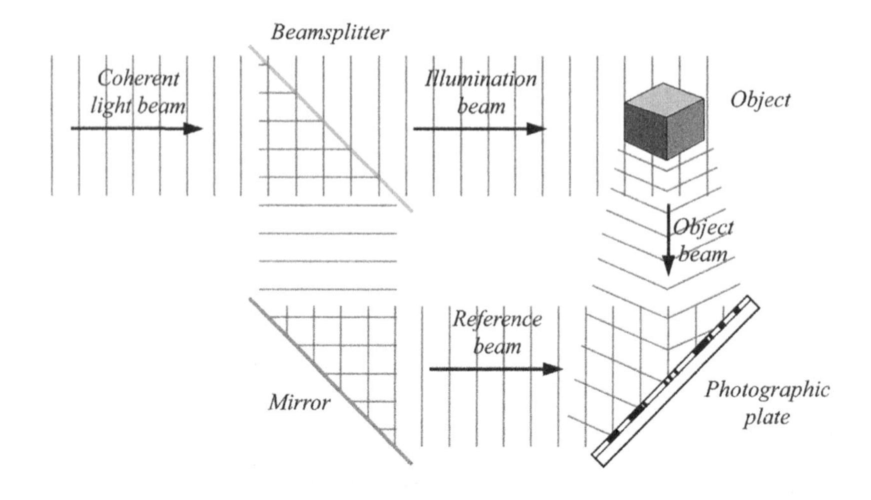

Figure 2.1 *A hologram recording process.*

From what follows, we learn that one of the most important properties of a hologram is that information about the whole system is equally distributed in each part of the scheme. Mathematical and experimental results have shown that in holographic systems—even if a part of a holographic plate is broken—each part will ultimately display the information and characteristics of the whole. The whole is in the parts, and each part contains the whole. The whole and the parts are just the same as one unified unbroken whole. When re-illuminated with one of the first coherent light sources, a three-dimensional image appears. Holographic systems can also generate three-dimensional virtual images. In fact, viewing a hologram is no different to viewing the actual object, which it represents because it contains all the information about the object (both amplitude and phase).

To sum up its meaning in optics, I would like to add that a hologram is a permanent record of the interference between two waves of coherent light. Experiment has shown that each part of the hologram contains each part of the interfering waves; each part of the hologram contains the entire image. The entire hologram contains more details about the image. However, the image is present in every part of the hologram.

Let us now look at its meaning in cosmology, QG, and ST. According to the SM, we live in four space-time dimensions, three dimensions of space plus one dimension of time. Also, our world is enveloped by the force of gravity.

What does holography say about our world? It suggests that:

The physics of space-time and gravity has an equivalent description in three space-time dimensions, without gravity.

What holography says is that all physics done inside a four-dimensional region of space-time can be equally described by a theory without gravity in a three-space-time region living on the boundary of the region. This boundary is simply called the hologram. The essence of the HP is that our world seems to behave like a hologram. In a nutshell, in a hologram, information about a three-dimensional object is stored on a two-dimensional surface, and this idea appears to be true for our world also.

In 1971 Hawking proved an important theorem that led to the development of the HP and settled the question of BH entropy. The theorem Hawking studied and finally proved is known as the Area Theorem of BH physics.

The Area Theorem of BH physics states that:

The area of a black hole horizon never decreases.

As we can see, there is a similarity with the second law of thermodynamics that says that:

The entropy of a closed system never decreases.

Is there a similarity between the two? Is it simply a coincidence? Are these two relationships equivalent? The answers to these questions had to wait for a young scientist named Bekenstein[18], then a young graduate student

---

[18] The connection between a BH and the laws of thermodynamics became apparent in the early 1970s. Work was led by Hawking and extended by Bekenstein. The latter found that the laws of BH dynamics are similar to the laws of thermodynamics. The work of Bekenstein, Hawking, and others has led to concrete realizations of many of these ideas, so that it becomes possible to associate thermodynamic properties with the physics

at Princeton University. In 1971 he was struck by the similarity between the area theorem and the second law of thermodynamics[19]. This similarity led him to argue in 1973 that a BH must have an entropy proportional to its horizon area. The idea was met with widespread strong disapproval from most cosmologists as the overwhelming majority argued that:

If a black hole has an entropy, it must also have a temperature, and it must radiate like a blackbody does at this temperature. This is impossible.

The reply by Bekenstein[20] was that:

"...if a black hole does not have an entropy the second law of thermodynamics will be violated. We could throw some matter into a black hole and lose entropy."

In the same year a British theoretical physicist named Stephen Hawking proved that BHs do radiate just like blackbodies, and he calculated the temperature of BH radiation. Moreover, the entropy of a BH is proportional to the horizon area. The work of both researchers has taught us that there are analogies between the theories of BHs and thermodynamics[21].

We have shown the special relationship that exists between the concept of holography and physics. Let us now retrace and introduce this idea in general terms, to gain a better understanding of it. We discuss some true motivations provided by study of the brain, microscopes, radar development, BHs, and entanglement. Records show that earlier work in biological holography may have started with some experiments in the 1920s led by a

---

of BHs and bring together in the same approach gravitation theory, quantum theory, and entropy or thermodynamics. It gave us a new insight into the thermodynamics of gravity.

19 Bekenstein, J. D. 1974. "Black-Hole Thermodynamics." *Physics Today* 33:24–31.

20 Bekenstein, J. D. 1974. "Generalized Second Law of Thermodynamics in Black-Hole Physics." *Physical Review* D 9 (12):3292–3300.

21 There are similarities between the behavior of BH area and entropy, namely that both quantities never decrease, but tend to increase irreversibly. Hawking showed that the BH area cannot decrease in any process, which is analogous to the fact that the entropy of a closed thermodynamic system cannot decrease. Bekenstein was to be proved right; the second law of thermodynamic is valid when the entropy of BHs is included.

brain scientist named Karl Lashley[22]. He discovered that regardless of which portion of a rat's brain he removed, the process was unable to eradicate its memory or ability to perform complex tasks learned before surgery.

The main outcome of Lashley's experimental work on rats showed that memories are not confined to a single, specific location, but are dispersed throughout the brain. This discovery was a novelty, and no real explanation for it was put forward. It took many years before Pribham resolved the puzzle when he made an analogy with the concept of holography—he found that memories are in patterns of nerve impulses rather than in neurons themselves, as had been previously believed.

Other records show that the possible origin of the theory of holography can be traced back to 1947–1948, in the work of the Hungarian scientist Dennis Gabor[23], who at the time was trying to improve the resolution of the electron microscope. In 1948 he put forward the idea of storing the full amplitude and phase information of a wavefront reflected from an object. This idea was based on Gabriel Lippmann's color photography. He applied a Fourier transformation to the creation of the hologram and later showed how the interference pattern could be successfully used to recreate any virtual image of the object simply by the application of the inverse Fourier process.

This led to the realization that from a simple parameter such as frequency, objects in space-time could be rebuilt in virtual form. For this remarkable work— invention and development of the holographic method—he was awarded the 1971 Nobel Prize. Not much happened after the work of Gabor until the mid 1950s. In 1956 Emerett Leith did some advanced work and reinvented holography as part of a top-secret high-resolution radar-imaging project. Unfortunately, no disclosure of the work he completed is available.

At or around the same time, records indicate that a Russian scientist named Yuri Nicholaevitch Denisyuk also did some experimental work in the field

---

22 Beach, Frank. A. 1961. *Karl Spencer Lashley 1890–1958 A Biographical Memoir.* National Academy of Sciences Washington D.C. Available at: http://www.nasonline.org/publications/ biographical-memoirs/memoir-pdfs/lashley-karl.pdf

23 Gabor, D. 1948. "A New Microscopic Principle." *Nature* 161:777–778.

of holography. From 1964 onward, advances in holography were made, all based to some extent on the principle of Fourier-transform holography first proposed by George W. Stroke and later extended by other researchers. The origin of holography in cosmology can be traced back to the 1970s in the study of BHs led by Hawking and Bekenstein. Following the work of Bekenstein,'t Hooft applied this concept in QG and Susskind emphasized its meaning in ST.

In 1971, Pribham applied the holography concept to neurophysiology and demonstrated in some experimental works looking at how the brain encodes information like a hologram, i.e. process information holography. Two years later the holographic concept of reality was suggested and extended by Miller, Webb, and Dickson.

In 1980 Bohm[24] put forward the theory of the implicate and explicate order that emphasizes the importance of the HP. With the publication of his book, the holography concept was firmly established. He developed a theory that many believe answered deep fundamental questions raised by Aspect's experiment. Bohm's ideas are elaborated on in his 1981 book, *Wholeness and the Implicate Order*. Along with many other scientists, Bohm argued that Aspect's findings imply that objective reality does not exist and that the universe is a gigantic and splendidly detailed hologram.

In the implicate order any element contains enfolded within itself the totality of the universe. This way of looking at the universe perhaps justifies Bohm's conviction that the correlation between particles was simply an illusion. He suggested that what was going on was that "at some deeper level of reality such particles are not individual entities, but are actually extensions of the same fundamental something".

In 1981 Jacob D. Bekenstein showed that the maximum amount of information that can be contained in any volume of space is limited, not by the volume of the space, but rather by the surface area that encloses that space. This discovery is known as the Bekenstein bound or the universal entropy

[24] Bohm, David. 1981. *Wholeness and Implicate Order*. Routledge and Kegan Paul: London.

bound. It defines and limits the amount of information on the holographic surface of the universe. A year later, in 1982, Ken Wilber took David Bohm's ideas seriously. This is expressed in his book, *The Holographic Paradigm*, where he commented that [quoted here from Gloria Alvino[25]]:

> *"For several years, those interested in human consciousness have been speaking wistfully of the 'emerging paradigm,' an integral theory that would catch all the wonderful wildlife of science and spirit. Here, at last, is a theory that marries biology to physics in an open system: the paradoxical borderless paradigm that our schizophrenic science has been crying for. It is appropriate that this radical, satisfying paradigm has emerged from Pribham, a brain researcher-neurosurgeon who was a friend of the Western Zen teacher Alan Watts, and Bohm, a theoretical physicist, and the close friend of Krishnamurti and former associate of Einstein."*
>
> —ALVINO 1996

In 1991 Michael Talbot[26], inspired by the work of Pribham and Bohm, argued that the universe can be understood as a kind of hologram. He discussed and addressed fundamental issues from the brain, the universe, space, and time from a holographic viewpoint, as well as the implications for a holographic theory of the universe. Many of the issues discussed here are well illustrated in his book *The Universe as a Hologram*. In the same year, Pribham expanded his earlier ideas. He was particularly concerned about the location of memories in the brain. Earlier experiments conducted by Karl Lashley showed that memories are not confined to a single, specific location, but rather are found dispersed throughout the brain. Pribham came to resolve the puzzle when he made an analogy with the concept of holography; he realized that memories are encoded in patterns of nerve impulses rather than in neurons as previously believed. His finding persuaded him and others working in the

---

[25] Alvino. G. 1996. "The Human Energy Field in Relation to Science, Consciousness, and Health." Accessed April 12, 2017. http://www.vxm.com/21R.54.html

[26] Talbot, M. 1992. *Holographic Universe*. Harper Perennial: New York.

neurophysiology community that the brain somehow uses HPs to perform its operations.

In 2000 Raphael Bousso[27] extended the HP as a universal law that holds for all surfaces, open or closed, regardless of location or shape. The many examples given by Bousso to illustrate the HP from a theoretical physics viewpoint are worth investigating. Seven years later Morgan showed the link between the HP and the quantum wave function. The exciting work of Morgan demonstrates that local interactions between quanta of mass may have immediate and non-local effects on the wave function throughout the universe.

In the next section we discuss Bohm's holographic ideas and the holographic world of Denis Gabor, Gerard 't Hooft, Leonard Susskind, Jacob Bekenstein, Raphael Bousso, and Karl Pribham. We elaborate on the evidence in support of it and its successes and weaknesses.

HOLOGRAPHY AND QUANTUM THEORY

David Bohm's Holographic Universe

As a respected theoretical physicist, Bohm has inspired—and continues to inspire—researchers and ordinary people all over the world. His immense and colossal contributions to knowledge encompass areas of science, religion, philosophy, peace, and dialogue, etc. Bohm's contributions to science are deeply profound and have yet to be fully recognized and integrated into the quest for the ToE. Throughout his long and illustrious career as a physicist, he has challenged the scientific orthodoxy.

Bohm is remembered by those who have encountered his writings and ideas as one of the most brilliant and distinguished theoretical physicists of his generation. The influences of his work extend far beyond theoretical physics. He was born in Wilkes-Barre, Pennsylvania, U.S.A., on December 20, 1917. He passed away in the autumn of 1992—struck down by a heart attack in front

---

[27] Bousso, R. 2002. "The Holographic Principle." *Reviews of Modern Physics* 74:825–874.

of his house, just a few days after finishing his last book with Basil Hiley, his long-term colleague at Birkbeck.

The last few minutes before Bohm's death are retraced in *The Jewish Quarterly* of autumn 2003. Naomi Gryn[28] recalls her last conversation with his wife:

> *On the last day, Dave said, "I really need to go in; we need to use the computer." He phoned me up, and he was like a young boy. It was years since I'd heard him with so much energy and so happy. He said, "We've finished, we've finished!" and then "I think I'm on the edge of something." Bohm took a taxi home from the station and collapsed outside their home.*
>
> —GRYN 2003, 96

A moving tribute by one of Bohm's close friends, Renee Weber, on the day he died stresses the importance of this great scientist. Quoted here from Scientix[29]:

> *"His Physics and Cosmology were so profound, so all-encompassing and so far ahead of his century that few people understood how creative his ideas were. Mainstream Physicists considered them too mystical, and few mystics could follow his subtle scientific reasoning."*
>
> —SCIENTIX 2016, 1

His career as a physicist started at the University of California (Berkeley). He later moved to Princeton's Institute for Advanced Studies. In 1957, he crossed the Atlantic to Britain where he worked at Bristol University until 1961. When the chair in theoretical physics opened at Birkbeck College in London, it was

---

[28] Gryn, N. 2003. "David Bohm and Group Dialogue—or The Interconnectedness of Everything." *The Jewish Quarterly*, Autumn 2003, pp. 93–97.

[29] Scientix. 2016. "DAVID BOHM, GENIUS OF MODERN SCIENCE IN SERCH OF TRUTH Explicate, Implicate Orders in Nature. Every Part of the Whole contains the Whole." Accessed February 18, 2016. http://tenets.zoroastrianism.com/DavidBohm GeniusofModernScienceinSearchofTruth.pdf

awarded to Bohm. He worked there as a professor of theoretical physics until he retired in 1983.

Bohm's father was from Munkacs in Israel, but had left for America when he was 16. Bohm's interest in science was noticeable at an early age when, as a young boy, he invented a dripless teapot. His father, then a businessman who ran a second-hand furniture store, urged him to make a profit on his invention. Surely this was an excellent opportunity to earn money during the depression in Wilkes-Barre, a poor mining town in Pennsylvania. His interest waned when he became aware that the first step in selling his product was to conduct a door-to-door survey to test the market demand; he gave up on the idea of starting a business and persuaded his father to let him study physics.

In the 1930s, he went on to study physics at Pennsylvania State College, where he became fascinated by quantum physics. After graduating from Pennsylvania, he went on to further studies at the University of California, Berkeley. He received his doctorate in 1943 and worked at the Lawrence Radiation Laboratory. He embarked on groundbreaking research in plasma physics while at Berkeley. As a graduate student, he discovered a (particular) collective movement of electrons in the plasma which is now called Bohm-diffusion. This behavior of electrons inside a plasma had a remarkable impact on the young Bohm, and was perhaps the inspiration for his later theory of *Wholeness and the Implicate Order*. Bohm remarked that once electrons were in plasma, they stopped behaving as individuals, but were rather acting as if they were part of a larger organized interconnected whole. This was his first of many contributions as a physicist.

In 1947 Bohm moved to Princeton University where he worked as an assistant professor. There, he extended his earlier research to the study of electrons in metals and was quick to notice similarities between the behavior of electrons in plasma and metals. The same effect that electrons had on him in earlier plasma research seems to repeat itself. Bohm remarked that the movement of individual electrons in metals appeared to produce a highly organized overall effect and that individual electrons could self-organize and create a sense of being alive. It was while at Princeton University that he met Albert Einstein. In 1950, he completed the first of his six books, entitled *Quantum Theory*, in which he reformulated the paradox of Albert Einstein, Boris Podolsky, and Nathan Rosen (EPR)[30]. Bohm's version of the EPR analysis involves components of spin in place of position and momentum.

In the early 1950s, Bohm, encouraged by Oppenheimer's wife and along with other students, joined the Communist party. He was deeply attracted to and interested in the philosophical and political ideas, but he gave up his membership as he found the meetings boring. As a student of Oppenheimer, he worked with him on the Manhattan Project to build the first atomic bomb in America. Oppenheimer was a scientific director of the Manhattan project and later chairman of the US Atomic Energy Commission; he voiced his concerns and reservations about the hydrogen bomb project strongly. The army was not pleased at all and was out to get him. Because of Bohm's connection with the Communist party, they tried to use him to get something on Oppenheimer. Bohm's days at Princeton were to be cut short as he was asked to betray his colleagues in 1951. As a man of principle and strong moral conviction, he refused to testify against them before the House Un-American Activities Committee, an act which led to his banishment from Princeton and all American universities. Oppenheimer advised him to leave America, so he headed to Sao Paulo in Brazil where he taught physics. He disliked the climate there as well the activities of many of the former Nazis who made it their home after the end of the war.

---

[30] In terms of implications and interpretation, QM remained disturbing even to its founders. Einstein himself believed that QM was incomplete. Some of the leading founders expressed the discomfort they experienced with the emerging orthodox dogma of the Copenhagen interpretation. It may be said—pessimistically—that no satisfactory interpretation of QM exists today. Einstein in particular did not like some aspects of the Copenhagen interpretation such as the loss of determinism, classical causality, the introduction of probability, and the consequent incomplete description of nature. Einstein held the view that there should be a mechanism that can describe phenomena more realistically, for example, by means of 'hidden variables'. He produced many objections to the theory and argued that it was incomplete. This led him to formulate various thought experiments, such as the famous EPR (Einstein-Podolsky-Rosen) paradox, to distinguish between a causal and a non-causal interpretation of the theory. He also made various attempts to refute the uncertainty principle.

Bohm was considered by Einstein to be his intellectual son and was among the pioneers who revolutionized quantum physics. Bohm's vision of reality argues for a multidimensional model of reality that treats the whole of existence, the universe, including matter, mind, and consciousness, as an entire whole. He continually argued for a dialogue but was frustrated and disturbed by the failure of communication between fellow scientists, particularly Einstein and Niels Bohr. He later went on construct an alternative to the Copenhagen interpretation of QT, called Bohmian mechanics. In his book *Quantum Theory*, he presented an account of the orthodox Copenhagen interpretation of quantum physics, which was formulated by Niels Bohr, Werner Heisenberg, and others in the 1920s. Bohm had a lot of objections and concerns about some aspects of their interpretation. For instance, he had reservations about accepting that subatomic particles had no objective existence and the fact that the quantum world was characterized by indeterminism and chance. He came to the realization that there might be deeper causes for the behavior of subatomic particles.

The objections that Bohm had were extended, and in 1952 he published two important papers on what came to be known as the causal interpretation of QT. In 1955 he was offered the chair of physics at Haifa Technion in Israel where he met his future wife Saral Woolfson. They fell in love at a party organized to welcome him and were eventually married. His days in Israel were numbered however, because the following year he accepted a teaching position at Bristol University in England.

In 1957 Bohm and Aharonov made a remarkable discovery, focusing on non-locality and the possible incompleteness of the quantum description. They demonstrated the existence of a "rather strange kind of correlations in the properties of distant things." Two years later, in 1959 and while at Bristol University, he published a paper on non-locality with Aharonov[31]. The result was the Aharonov-Bohm effect, which shows how a magnetic field can influence the behavior of electrons confined far away from the field. In a nutshell,

[31] Aharonov, Y. and Bohm, D. 1959. "Significance of Electromagnetic Potentials in the Quantum Theory." *Physical Review* 115:485–491.

they found that under some conditions electrons can feel the presence of a nearby magnetic field even if they are traveling in regions of space where the field strength is null. The Aharonov-Bohm effect is perhaps one of the best examples that demonstrates the existence of quantum interconnectedness.

Bohm's Concerns on the Incompleteness of Quantum Theory

Throughout his illustrious career as a physicist, Bohm pointed out the incompleteness of QT on many occasions. His views are well illustrated and highlighted in his books, particularly the concern that our two pillars of modern physics, QM and relativity theory, contradict each other. This contradiction, he argued, is not only minor in detail but truly fundamental. He put forward the argument that our new physics is self-contradictory at its foundation, and does not answer many questions. He also pointed out the lack of interest shown by many physicists in what he considered a significant discrepancy. It was his quest to solve this dilemma which led him to the discovery of his theory of undivided wholeness.

In his book *Quantum Theory* Bohm presented a reformulation of the EPR experiment. Bohm's version of the EPR analysis involves components of spin in place of position and momentum, focusing particularly on non-locality and the incompleteness of the quantum description. He highlighted the problem of the inseparability of the observer and the observed. These were some of the challenges in QM that had occupied him throughout his life. Bohm, like Einstein, shared a belief in the incompleteness of QT; they were both dissatisfied with the conventional approach. Their faith was elucidated in their many discussions. Even before the publication of his book, he had some doubts about many assumptions underlying the Copenhagen interpretation of quantum physics.

He found it particularly difficult to accept that subatomic particles had no objective existence and took on certain properties only when an observer tried to observe and measure them. He also rejected the notion of indeterminism and chance, widely taught in the conventional approach. These two dilemmas, and possibly others, led him to suspect that there were perhaps deeper causes for the behavior and nature of the subatomic world. In 1952, a year after the publication of his book, he developed the view that subatomic particles are highly complex and dynamic entities, contrary to the prevailing view that subatomic particles such as electrons were simple and structureless particles.

He struggled with the view that the motion of subatomic particle was uncertain; he argued that they follow a precise path, which is determined first by conventional physical forces, but then also by a subtler force known as the *quantum potential*. It is this force that guides the motion of particles by providing active information about the whole environment. He went on to argue that the quantum potential pervades all of space and provides direct connections between quantum systems. Bohm's work received a boost after the Aspect experiment to test quantum interconnectedness was performed in 1982. It showed beyond any doubt that subatomic particles that are far apart, regardless of the distance between them, can communicate in ways that cannot be explained by the transfer of signals at or slower than the speed of light.

All this life Bohm wrestled with fundamental questions raised by physics. He tried to explain the non-locality which the Alain experiment supports and his concerns have resurfaced once more in the search for the ToE. It is also worth pointing out that his human and sensitive side, his care for all beings, was demonstrated by his pleas for dialogue among people. He was particularly concerned about the lack of dialogue between his fellow scientists, especially Albert Einstein and Niels Bohr, as well as damage to the environment and ecology, the divisions between races, and the failure of dialogue between nations that led to fragmentation and isolation.

Wholeness and the Implicate Order

In the 1980s Bohm developed his theory of the *Implicate Order* to explain the bizarre behavior of subatomic particles—well illustrated in the EPR experiment. He was convinced that the behavior of subatomic particles might be caused by some hidden means, and this 'hiddenness' might be a reflection of a deeper dimension of reality. This realization led him to believe that space and time might be derived from this deeper level of objective reality that he calls the implicate order, and that within this implicate order, everything is connected. Bohm argued that the universe is like a hologram. He asserted that what we see around us, the universe, is a kind of illusion just like a holographic image. He suggested that there is a deeper order of existence that gives birth to all things that exist in our physical world, in the same way that a piece of holographic film gives birth to a hologram. This deeper order of existence or deeper level of reality is the *implicate order* (i.e., enfolded or hidden). It is in contrast to our visible physical existence, the unfolded or *explicate order*.

Let us explore what Bohm means when he says that the universe is like a hologram. This is better explained by using the concept of holography taken from optics, on which we have elaborated throughout. A piece of holographic film generates an image. The film that generates the image is an implicate order, merely because the picture encoded in its interference patterns is a hidden totality (hidden from us). The hologram projected from the film is an explicate order as it represents the perceptible visible version of the image. A hologram records details down to the wavelength of light. It is similar to dense information storage. This explains how information about the entire holographic scene is enfolded into every part of the film. It echos the implicate order in the sense that every point on the film is completely determined by the overall configuration of the interference patterns. Even a tiny part of the holographic film will reveal the unfolded form of an entire three-dimensional object.

Holography, derived from the Greek, means complete; in writing it means that every part of 'the writing' contains information about the whole. A hologram, which is the material manifestation of a holograph, is a three-dimensional image formed by the interference of light beams from a coherent light source. When the hologram is illuminated, a realistic three-dimensional image is produced, e.g., a photograph of the interference pattern. In a hologram, each part contains some information about the whole; the information or features are not localized, but are equally distributed in each region. What is the process of holographic recording and decoding? We can understand holography through the phenomena which constitute it, namely interference and diffraction, both based on the wave theory of light. Generally speaking, the hologram is recorded on a flat surface and contains the information about the three-dimensional wave field. This information is coded in the form of interference stripes, invisible to our eye due to the high spatial frequencies.

Although the hologram is recorded on a photographic plate or a flat piece of film, the idea behind the hologram is different to the recording of an image in the conventional sense of photography. In conventional photography, what is recorded is only the intensity distribution in the original scene, rather than the phase and amplitude of the light waves coming from an object. Furthermore, all information about the optical paths is not recorded. Photography produces two-dimensional picture of the object. Whereas in hologram generation techniques, the recording process depends both on intensity as well as phase of light wave. Hence, the phase information is converted into variations of intensity and this is achieved by using coherent illumination and adding a reference beam derived from the same source. The intensity at any point in the interference pattern also depends on the phase of the object wave; as a result, the hologram contains information about both the phase and the amplitude of the object wave, It gives three-dimensional picture of the object.

The intensity is the only parameter of light which sensors are directly responsive to, e.g., eyes or photodiodes. Each optical wave field is made up of an amplitude distribution and a phase distribution. Let us stress that detectors or recording materials such as photographic film only register intensities, meaning that the phase is lost in the registration or recording process. In holography, the phase information is coded by interference into a recordable intensity. For instance, a common vision sensor encodes illumination intensity into voltage, current, or timing information; all the information is encoded, both amplitude and phase, on arrival at the sensory apparatus. We stress that holography uses interference patterns—it involves the transformation of light and optical information. In wave mechanics terms, the coded recording of a wave is called a hologram, thus holography is based on the wave nature of light. No wonder Bohm[32] suggested, in 1992, that matter is revealed as an ocean of both energy and light:

> *"Matter, as it were, is condensed or frozen light...all matter is a condensation of light into patterns moving back and forth at average speeds which are less than the speed of light...It is energy and it's also information – content, form and structure. It is the potential for everything."*
>
> —RINPOCHE 2008, 279

Bohm thus proposed a new order—the implicate order, where "everything is enfolded into everything." In contrast, in the explicate order, things are unfolded; the manifest world is what Bohm refers to as the "explicate order". It is secondary, it flows out of the law of the Implicate Order. *In Wholeness and the Implicate Order,* Bohm comments that we live in a multidimensional world; one of the multidimensional worlds is our three-dimensional world, namely objects, space, and time, which he calls the explicate order. This is the level on which most physics operates today, with findings presented in the form of equations, the meaning of which is unclear to most people. Symbols represent these equations, and there is no connection between the mathematical equation and the actual physical reality. This explicate order is what can be perceived by our senses or instruments; this is what is in the domain of physical reality—for instance, rocks, stars, moons, planets, ourselves, etc.

All these separate objects, entities, and structures are derived from a deeper implicate order of unbroken wholeness. This implicate order is therefore the all-encompassing background to our daily experiences, including physical, psychological, and spiritual. Beyond the implicate order, there is the super-implicate order. According to Bohm, the observable universe of our experience is merely a reflection or extension of higher mystical universes.

---

[32] Rinpoche, S. 2008. *The Tibetan Book of Living and Dying: A Spiritual Classic from One of the Foremost Interpreters of Tibetan Buddhism to the West.* Random House: London

Secondly, our consciousness is confined to the three dimensions of space (length, breadth, and height), and one dimension of time. Lastly, the higher universes or orders have n dimensions where n can be any number, even infinity. Beyond the super-implicate, there are many other orders merging into an infinite n-dimensional source.

Bohm saw everything around or beyond us as an interconnected whole; matter and events interact with one another, and time, space, and distance are simply an illusion. His holographic model of the universe shows that the whole can be found in the minutest part: time, space, and distance, present, past, and future, all occupy the same non-space and non-time. According to Bohm, the subatomic quantum is conscious and everything around us is aware, including planets and inanimate objects. The fundamental concepts of Bohm's theory are contained in the notion of holo-movement. Consciousness, according to Bohm, is an integral feature of holo-movement, in which mind and matter are interdependent. Bohm considers three dimensions of reality: explicate, implicate, and super-implicate orders. The explicate order is manifest and includes the three-dimensional world of objects, space, and time. The implicate order is the deeper, hidden, enfolded one which includes all background to our experience; i.e., physical, psychological, and spiritual. Lastly, Bohm's other dimension is a subtler dimension called the super-implicate order.

Modern science, particularly quantum physics, supports the ideas of interconnection and interdependence; the physical object seems to be independent and stable. However, we learn from physics that this is not exactly the case. Quantum physics teaches us that matter and energy are not separate entities. Energy flows in waves that form patterns. A stone, for instance, is simply a pocket of reality where the energy is denser. A subatomic particle is not a little dot of matter that can be examined; it is rather a dancing point of energy. These particles cannot be understood as separate units. Physicists can describe subatomic particles only by analyzing how they interact with one another. We have learnt that to describe subatomic particles meaningfully is to explain the way they interconnect, and that quantum physics shows that all reality is an integrated web of energy. At the subatomic level, life is interconnected.

According to ST, the particle is not point-like but is made up of a tiny loop. Each particle contains a vibrating dancing string. Everything is identical but each string can vibrate, and each one has a different vibration: the note the string creates determines the kind of particle it will be. Accordingly, these vibrating strings make up everything in the universe, such as all physical matter and forces. On the other hand, the SM informs us that space-time consists of three dimensions of space (length, width, and depth) plus the dimension of time. All objects and events exist in these four dimensions. In contrast, according to ST space-time can have up to nine dimensions of space plus one dimension of time. What we learn from ST is that matter and energy are the same and that, from the smallest particles to the largest solar system, they obey the same principles. At virtually all levels life is interconnected. ST thus shows the interconnectedness of life at all levels, large and small.

We also learn from chaos theory that complex systems in nature have an underlying order, even if they are chaotic and virtually unpredictable. For example, the flapping of a butterfly's wings at one location in the world may affect the weather on the other side of the Earth. The realization is that any action, no matter how insignificant, can affect everything else. For instance, the success of a small or big company depends on each staff member. Likewise, the success of each member depends on the success of the enterprise. Similarly, the success of each living being depends on the success of the community. Even though the universe appears to be made up of individual parts, and these parts may function as independent entities, the parts are all made of the same energy and are connected to form one giant whole. So, each part is not separated and isolated from the others even though it may function by itself. Bohm built his theory on the idea that reality unfolds from one original infinite source. He suggested that the information of the entire universe is contained in each of its parts. The explicate order is the separate parts of the world that we see. On the other hand, the implicate order enfolds all these parts into one whole.

The implicate order is the original energy, the source of all reality. This source is called the subtle nonmanifest. The later gave rise to all space and time, all dimensions, and all planes of existence. Accordingly, all beings are born from this source, are connected, and share consciousness. They go through a process, have new experiences, and grow in wisdom and knowledge. They then unfold back into this source. Reading *Wholeness and the Implicate Order* was eye opening for me. I came to the conclusion that to understand quantum physics one must think regarding unbroken wholeness—everything in the universe is infinitely and internally connected to everything else.

Bohm was also strongly influenced by spirituality—his spiritual ideas somehow appear to shape his physics. It is possible that Bohm did not only want to create a new physics, but an entire world-view beyond physics. This can be seen in his writings and views on ecology, politics, peace, and dialogue. However, here we want to stress that Bohm and Basil Hiley proposed a description of physical reality in which wholeness is the foundation of physical theories. This is perhaps due to the fact that several myriad problems which have arisen when attempting to form a consistent and unified description of reality through physics have not been successful tackled.

The incompatibilities between the two most physical theories, QM and the theory of general relativity, have led to some physicists proposing wholeness as a theory that includes both. The reductionist approach to physics, in which we treat parts as a whole, has been successful over many centuries. In the reductionist approach, we treat our world as if it is made up of a separate entity that interacts with one another. The discovery of QT led us to realize that the reductionist approach would not work. Early quantum physicists discovered that the QT was not reductionist. This is well illustrated in the book *Atomic Theory and the Description of Nature*, published in 1934, where Niels Bohr emphasized the role wholeness plays in QT.

To emphasize the role of wholeness, consider two quantum systems that interact. Let us call them system I and system II. Once they interact, the system of interaction is made up of one inseparable system and can no longer be regarded as being made up of two separate parts. Also, the wave function of the interacting systems cannot now be expressed as a product of the wave functions of the two separated systems; instead, an inseparable totality is formed. Personally, I do not see how a reductionist approach can be supported or maintained. In the *Undivided Universe*, Bohm and Hiley propose an ontological interpretation of QT with various other examples of how a non-local form of wholeness plays a central role in QT. Many other physicists have proposed similar ideas.

An article published by Barbara Piechociska[33] in 2008 is a brave attempt to extend Bohm and Hiley's work by proposing a mathematical framework within which a physical theory based on wholeness may be expressed. Piechociska shows step-by-step how features of wholeness, such as wholeness being indescribable, give rise to a self-similar or holographic type of order which is reflected in the mathematics. Furthermore, she shows how all the sets in the mathematical universe may be expressed as emerging from the dynamics of wholeness and how this mathematics may be further developed to connect with a physical interpretation.

We have elaborated in detail about wholeness and the reductionist approaches. As we are nearing the end of this part of the book, it is time to define both concepts and look closely at their meanings. Several authors have addressed the elusive nature of wholeness. However, the first step to understanding it is to highlight some of its features. According to Piechociska, wholeness is viewed as the totality that implicitly unites and gives rise to everything observable, but it is more than the observable. On the other hand, as stressed by Bohm and others, the reductionist approach of working with parts is an inconsistent picture with which to understand the physically measurable reality; working with parts to achieve wholeness presents us with an insufficient and limited picture of wholeness, and is perhaps misleading.

The description regarding parts is what matters from the reductionist approach, contrary to the wholeness approach which includes the holistic

---

[33] Piechocinska, B. 2004. "Wholeness as a Conceptual Foundation of Physical Theories." [Online] Accessed May 06, 2010. http: arXiv: physics/0409092v1 [physics.gen-ph.]

perspective. Thinking in terms of parts may not be the correct approach to describe physical reality. The reductionism approach views the system it studies as being made up of separate identifiable parts which interact with each other. The wholeness or holistic view stresses the indivisible wholeness as basic and fundamental, and assumes unity and totality. In support of wholeness, Barbara Piechociska writes:

> *"The holistic view assumes that there are no separate parts. Although there may be distinguishable modes on some level, on a deeper level we will see that separation and precise distinction are not possible. In other words, on that level we would not be able to find identifiable parts. As a consequence, we could not have both strict causality and strict locality. We could have one of them, but that would imply that we would not be able to have the other."*
>
> —PIECHOCINSKA 2004, 9

We now sum up Bohm's considerable contributions to physics. This conclusion, however, brings us to one fundamental question: what is the nature of matter and what it is made of? In 1992 he suggested that matter is revealed as an ocean of both energy and light. This is an interesting statement. Now we should ask ourselves what matter is made of. Where is the substance of matter? We are taught to believe that the substance of matter is in the atom. However, we learn from the same physics that atoms are made up of electrons, protons, and neutrons. Is the substance of the matter in electrons or protons? The answer is no, simply because these particles are made of further tiny particles called quarks and gluons. Let us then ask, is the substance of the matter in the quarks or gluons? We do not think so, simply because a quark and an antiquark can annihilate each other to produce photons, which are electromagnetic energy. So, as we can see, the attempt to persuade us that a stone is made of some solid entity is misleading and incorrect, hence questioning the whole validity of the SM of physics.

The SM of particle physics has been a success and its fundamental importance cannot be underestimated. However, it must be stressed that it has several shortcomings, which may suggest that it cannot be considered as a complete description of nature. Perhaps some of its successes should be embedded in a larger and more inclusive framework like Bohm's HU paradigm?

Finally, let us sum up Bohm's main ideas about the HU:

- Holography is the interconnection of all things.
- Evolution is informed by feedback loops.
- Matter, sound, and light are different vibrational frequencies of one universal energy.
- Everything is interconnected. That is why certain events appear to be connected, even though they do not interact with each other physically and are some distance apart.
- The holographic paradigm uses the hologram as a metaphor. A hologram contains all necessary information to generate an image of the entire hologram.
- Wholeness is viewed as the totality that implicitly unites and gives rise to everything observable, but it is more than the observable.
- The reductionist approach of working with parts is an inconsistent picture with which to understand the physically measurable reality. Working with parts to achieve wholeness presents us with an insufficient, and above all, limited picture of wholeness. It is perhaps misleading.
- The implicate order is the deeper, hidden, enfolded, and all-encompassing background to our daily experiences, including physical, psychological, and spiritual.
- The explicate order is manifest; it includes the three-dimensional world of objects, space, and time. This explicate order is what can be perceived by our senses or instruments; these are in the domain of physical reality. Examples include rocks, stars, moons, planets, and ourselves, etc.

HOLOGRAPHY AND QUANTUM GRAVITY

The Myth of Gravity

My eldest son reads a primary science book explaining gravity. He says to me "Papa, look, every time I jump in the air a force pulls me back down to Earth, do you know why?" Ironically, I said no. He went on to explain to me that it is simply because of gravity. Gravity pulls everything down to Earth and all masses attract each other, he added. Finally, he went on to ask me why gravity behaves in this way. I just said to him, son, this is just the way nature works. Newton, in response to a similar question, said "hypothesis non fingo." This translates as "I do not have a clue."

It is astonishing that such a simple and straightforward question about a common phenomenon we experience every minute of our lives has not found a direct answer for many centuries, even from the brightest physicists. Why is gravity there in the first place? What is gravity? Commonly, it is defined as an attractive force which all matter possesses. Matter attracts other matter and the strength of the attraction depends on two things, the mass of the objects and the distance between them.

Gravity and its derivative gravitation come from the Latin word gravitas, from gravis, i.e. heavy. It has been reported that the modern meaning of attraction did not appear until Newton's time. For scientists like Galileo and Copernicus, up to the beginning of the twentieth century, the word gravity had no concrete meaning. We learn from classical physics that gravitation or gravity is a natural phenomenon in which objects or bodies with mass attract one another. Gravitation is responsible for keeping our planet in its orbit around the Sun. It is also responsible for keeping the Moon in its orbit around the Earth.

Without gravitation we would witness chaos in the universe as it causes dispersed matter to coalesce. Most important of all, gravity is responsible for the formation of tides, for natural convection, etc. For instance, we learn that in space objects maintain their orbits because of the force of gravity acting upon them. Despite the successes of the SM of particle physics in unifying the three fundamental forces of nature, the fourth—namely gravity—remains elusive. This is due perhaps to our lack of understanding of its origin and why it is there in the first place. Among all natural forces that exist, gravity is the force we are most familiar with, but so far, its origin has eluded even the brightest physicists.

Newton's physics explains how an apple fall towards the Earth. Einstein went a step further by explaining gravity as the warping of the fabric of space-time. All of these theories so far have described how gravity works, but their shortfall remains in their inability to explain how it arises. For instance, Newton had a lot of reservations and doubts about the action of gravity at a distance. For centuries, this question among others has not been satisfactory answered. How can a massive object attract another at great separation without any mediation?

Another question that has also remained unanswered is how to explain the attraction between two objects without a time to action? Recently, many publications have pointed out that gravity might not be a fundamental force at all, but rather an emergent property of the deeper underlying structure of the universe. Since the publication in 2010 of Verlinde's[34] paper entitled "On the Origin of Gravity and the Laws of Newton", many follow-up papers have examined this idea in a new light. The overwhelming feeling that arises from those publications is that a new worldview emphasizing the primary role of information over matter and energy is likely to take center stage.

However, I would first like to look at the history and implications of the theory of gravity, from ancient Egyptian times to Einstein, as well as recent attempts by several researchers, who believe that the universe could be emergent and holographic. I will also examine how this new model, in which gravity may not be a fundamental force, could change our picture of space-time, giving a boost to the physics of computation and information—what we now call digital physics. I attempt in the following pages to present our understanding of how gravity came into existence, beginning with ancient

34 Verlinde, E. P. 2010. "On the Origin of Gravity and the Laws of Newton." Accessed July 07, 2017. http:// arXiv: 1001.0785[SPIRES].

times and ending with recent advances, including the work of Einstein who summed up all previous efforts in one elegant theory.

HISTORY OF GRAVITATIONAL THEORY

Earlier Ideas

The history of astronomy and gravity goes back several thousand years. There is no doubt that all ancient cultures and civilizations, from Africa, Asia, America, and Europe, etc., had theories about how the universe was created and how things around them worked and behaved. Newton, for instance, was the first to put together in a connected way simple laws which govern the movement of the planets and other objects when under the influence of gravity or any other force. He is widely regarded as the father of classical mechanics. Further, he established in an ingenious way what is known today as Newton's laws of motion[35]. These he demonstrated with simple experiments.

All civilizations tried to understand gravity. The early Egyptians believed that the universe resembles a large rectangular box with Egypt at the center of the bottom. As Egypt was a powerful nation at the time, countries that border it were heavily influenced by it and had the same concept of an enclosed space. There is an article by Patrick Tabaro[36], entitled "African Heritage Series: Blacks gave Science to Europe", in which there is an interesting quote by

35 It is well documented that many of the concepts developed in Europe were originally from Egypt. In the fifth or sixth century, the Muslims invaded North Africa and learnt from what was left in Egypt. They took it to a new height, so that in the eleventh century, Al-biruni was the first to introduce experimental scientific method into mechanics. Moreover, it seems that the first and second principles of what is known today as Newton's laws were in fact discovered by Ibn al-haytham and Avicenna. Newton's second law (law of constant acceleration) may also have been first discovered by Hibat Allah Abul-Barakat al Baghdaadi. Finally, other Muslim scholars such as Jafar Muhammad Ibn Musa, Ibn Shakin, Ibn al-Haytham, and Al-Khazimi, developed the theories of gravity.

36 Tabaro, J. P. 2011. "African Heritage Series: Blacks Gave Science to Europe." Accessed February 09, 2011. http://www.blackherbals.com/

Count Volney who accompanied Napoleon Bonaparte to Egypt (1799–1801). Volney was a scholar of great insight and objectivity, and went on to observe that original Egyptians:

> *"...Must have been real Negroes of the same species with all the natives of Africa..."*
>
> *Patrick Tabaro went on to comment that:*
>
> *"Volney was a French man, and so he had no desire or motive to falsify history in favour of Africans against Europeans. The Greeks and Europeans acquired their knowledge and science from Africans (Egyptians). To say this is to challenge the whole edifice of European claim (Some quarter's racists) to superiority over Africans, and this is only important to Africans who demand to be treated with dignity."*

The same school of thought is found and well illustrated in the writings of Martin Bernal[37] [quoted by Tabaro], author of *Black Athena, The Afro-Asiatic Roots of Classical Civilisation*, first published in 1991. He explicitly states (and argues throughout his book) that Greeks acquired their civilization from Egyptians and some Asiatic civilizations. His arguments are supported and demonstrated by many inventions and discoveries made by non-Europeans, for instance, printing, gunpowder, prismatic compact, etc.

The history of science that shows the contributions of earlier work by African scientists has been studied at some African universities, particularly in Senegal, Mali, and South Africa Unfortunately, in the past this has been inaccessible to the public and to historians of physics. However, recently with the availability of the internet, these records are now freely available. It has now become clear that many discoveries claimed by Kepler, Copernicus, and Newton were African and Asian in origin. We note that Kepler's theory of the movement of the planets was preceded by Egyptians. In addition, Copernicus's idea of a geocentric model was inspired by the Egyptians who

---

[37] Tabaro, J. P. 2008. "The Legacy of Ancient Egypt in Africa Today." Accessed February 9, 2011. http://onekamit.over-blog.com/tag/civilisations-verites et mensonges/.

worshipped the Sun because of its central position in our solar system. After studying the Egyptian system, Newton, in his book *Philosophiae Naturalis Principia Mathematica*, stresses that he was convinced that Egyptians knew about gravity and were acquainted with atomic theory.

For instance, the Egyptians were leaders in chemistry—the etymology of the word chemistry or *kemi*, simply means black people of Egypt. Furthermore, if we look at information and scripture on the Great pyramid, it easy to tell that ancient black Egyptians knew about the movement of the planets around the Sun. Egyptian science demonstrates that Africa became civilized 6,000 years before Europe, as shown by the invention of writing by Africans. In fact, Greek science and philosophy surfaced more than 2,500 years after Egyptians had mastered all ideas that led to European civilization. As such, there is no doubt that the civilization of the ancient Egyptians was very well advanced in science and technology. Perhaps the ability to understand and overcome gravity was the reason behind their progress and technological advancement.

The Egyptians may have developed technology to overcome gravity in order to lift and move heavy objects such as stones, to build pyramids, or to build other huge structures. Few people, even academics, are aware of (or tend simply to ignore) the immense contributions of Africans and the legacy of Africa to science, particularly physics and chemistry. One of the main reasons for this is the plundering of African resources during colonization, as well as the earlier development of anti-Africanization: the tendency of a group of intellectuals to regard Africa as a backward continent.

In Asia, particularly in India, efforts to understand and explain the phenomenon of gravity started from the eighth century BC. One of the earliest researchers (Kanada) believed that "weight causes falling; it is imperceptible and known by inference." Furthermore, the Indian philosopher Brahmagupta (AD 628) presented his thoughts on the subject of the heliocentric system by arguing that "all heavy things are attracted towards the center of the Earth" and that "all heavy things fall to the Earth by a law of nature, for it is the nature of the Earth to attract and to keep things, as it is the nature of water to flow, that of fire to burn, and that of the wind to set in motion, etc."

In the ninth century, Muhammad Ibn Musa wrote that there was a force of attraction between heavenly bodies in his work entitled *Astral Motion and the Force of Attraction*, thus anticipating Newton's law of universal gravitation. Al-Khazini, in the *Book of the Balance of Wisdom*, also made considerable efforts to understand and differentiate between the notions of force, mass, and weight.

Ancient Greeks

The history of gravity is inseparable from the progress made in astronomy since olden times; for instance, during the Middles Ages the geocentric astronomical model was the dominant school of thought. Most of the scholars at the time believed that the Earth was the center of the universe: the planets, the Sun, as well as other celestial bodies, move around the Earth. One of the models of the universe described by the Greeks put the Earth in the center of the universe, in a sphere in which the stars were fixed. The Greeks also believed that the Sun and the Moon were moving in circles around the planet Earth. There were many objects in the sky which they named planets, from the Greeks word for wanderer.

Most of the leading Greeks scientists and philosophers at the time represented the planets as circling the Earth, and all objects falling onto the Earth. Furthermore, they also believed in the existence of four natural elements (earth, water, fire, and air) and considered earth as the heaviest of all of these. As such, the Earth itself was thought to be in the center of the universe so that all objects will fall towards it. Plato educated the leading Greek philosopher and mathematician, Aristotle (384–322 BC), and he in turn learned from Socrates.

Aristotle developed his ideas from what he learned at Plato's Academy. For example, the Plato Academy taught students that the circle and the sphere were perfect shapes in two and three dimensions; the Earth was spherical, and that the Moon was at 60 Earth radii from the earth. They also believed that the Moon, the Sun, and the planets were embedded in crystalline spheres that revolved around the center of the universe, the Earth. Aristotle particularly believed that the material world consisted of four elements: earth, water, air, and fire. He argued that each element has a natural place in the universe, for e.g., earth belonged in the center, the air was above the water, water was in a layer covering the earth, and fire was in space, specifically, in the air.

Aristotle believed that each of the four elements had a natural tendency to return to its place so that, accordingly, heavy bodies should fall toward the center, i.e. Earth. He implied that heavier bodies or objects should fall faster than lighter ones. Aristotle's model of the universe included the Moon, the Sun, as well as the visible planets and the fixed stars. Celestial bodies circled Earth attached to nested ethereal spheres—his geocentric (Earth centered) model was dominant for many centuries until the Polish astronomer Nicolaus Copernicus came up with the Sun-centered or heliocentric model. Figure 2.2 shows Aristotle's geocentric model of the universe.

Another prominent Greek mathematician of the time, Aristarchus, came up with an interesting proposition: the Sun was the center of the universe and Earth revolved around it. He was led to this conclusion following his astronomical calculations which showed that the Sun was much larger than the Earth, and therefore, the small body should orbit the larger one. In the second century AD other researchers showed that Aristarchus's model could not be sustained, based as it was on spherical perfection. This conclusion was based on observations of Mars which revealed that the planet traveled across the sky.

In Africa (Egypt), in the second century AD, Claudius Ptolemy proposed a new model which kept some ideas of the geocentric model but suggested that Earth was still the center of the universe. The difference was that planets revolved on small circles called epicycles while the centers of the epicycles revolved on larger circles. Ptolemy put his ideas on a firm foundation by creating a powerful system of circles and epicycles that could explain all types of movements observed in the sky. This is well illustrated in his book *The Almagest*, which was a must-have reference astronomy book for over 1,400 years.

Let us now sum up Aristotle's ideas specifically about gravity. He believed in the law of cause-and-effect, i.e., there is no effect or connection without a cause. The reason heavy bodies or objects fell on Earth was related to their nature, which caused them to move downward toward the Earth; Earth was believed to be the center of the universe. Aristotle believed that heavy objects were not attracted to the center of the universe because of a force called gravity, but that downward movement toward the center of the universe is due to the inner gravitas or heaviness.

In contrast, light objects move—by their nature—upward toward the inner surface of the sphere of the Moon. The same line of reasoning is echoed and resonates with Vitruvius[38], who was a Roman engineer. In his book *VII of De Architectura*, he sums up his thoughts that gravity is not dependent on a substance's "weight" but rather on its "nature."

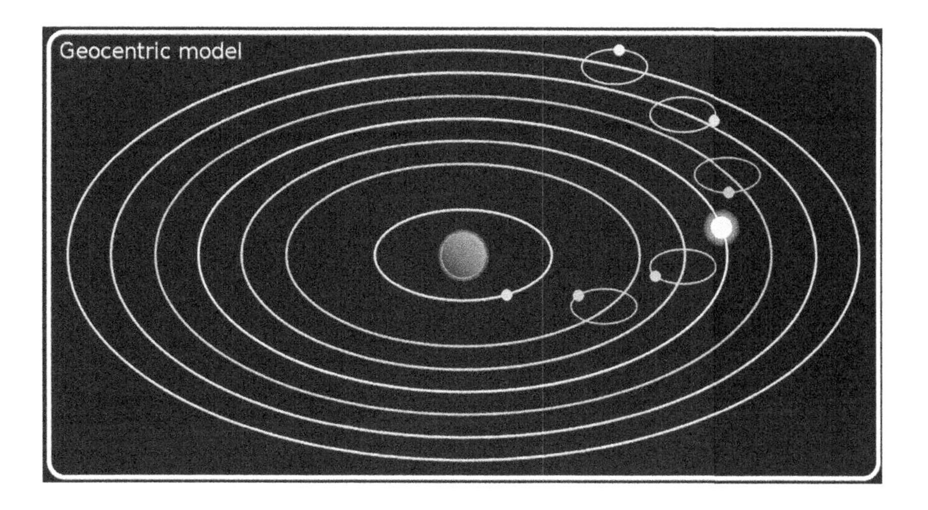

Figure 2.2 *Aristotle's geocentric model of the universe included the Moon, Sun, as well as the visible planets and the fixed stars. The celestial bodies circled the Earth (in blue) attached to nested ethereal spheres.*

[38] Vitruvius, M. P. 1914. "7". In Alfred A. Howard. *De Architectura libri decem* [Ten Books on Architecture]. VII. Herbert Langford Warren, Nelson Robinson (illus), Morris Hicky Morgan. Harvard University, Harvard University Press: Cambridge. p. 215.

"If the quicksilver is poured into a vessel, and a stone weighing one hundred pounds is laid upon it, the stone swims on the surface, and cannot depress the liquid, nor break through, nor separate it. If we remove the hundred-pound weight, and put on a scruple of gold, it will not swim, but will sink to the bottom of its own accord. Hence, it is undeniable that the gravity of a substance depends not on the amount of its weight, but on its nature."

—VITRUVIUS 1914, 215

Middles Ages

The geocentric astronomical model was dominant until Nicolaus Copernicus[39] (1473–1543) proposed his heliocentric model. The heliocentric or Sun-centered model places the Sun at the center of the universe, Figure 2.3, in contrast to the geocentric model. In Copernicus's model, the Sun was in the center of the universe, with all the planets orbiting it in perfect circles. Unfortunately, his model could not account for why the planets circled the Sun. His ideas were very revolutionary, putting him far ahead of his contemporaries. He published them in a book toward the end of his life entitled *On the Revolutions of the Celestial Spheres,* where he passionately argued and promoted the idea of the Sun being placed in the center of the universe instead of the Earth.

Copernicus's heliocentric model was widely opposed by the church and parts of society as it contradicted the dominant views of the church at the time. Fortunately, he received support from some of the great astronomers of the time, namely Galileo Galilei and Johannes Kepler. Then came Tycho Brahe (1546–1601), a Dane born three years after Copernicus passed away. At the age of 26 he discovered a new star and invented his cosmological model of the universe, which placed the Earth at the center of the universe. The

[39] Copernicus. N. 1543. *On the Revolutions of the Celestial Spheres* (edited by Jerzy Dobrzycki and translated with commentary by Edward Rosen). Polskiej Akademii Nauk: Wrocław, Poland

difference between his model and previous ones was that the Sun moves around the Earth, and the rest of the planets circle the Sun.

A further blow to the geocentric model (and support for the heliocentric one) came from the Johannes Kepler (1571–1630), a German mathematics school teacher. In 1596 he wrote his first book *Mysterium Cosmographicum*[40], which caught the eye of Tycho Brahe and led to a fruitful collaboration between the two men. Kepler calculated the orbit of the planets around the Sun using astronomical data from Tycho Brahe's work. There is no doubt that Kepler, using Brahe's data and analysis, took earlier work to new heights to produce what we know today as Kepler's laws. Kepler's laws state that: the orbit of every planet is an ellipse with the Sun at a focus; a line joining a planet and the Sun sweeps out equal areas during equal intervals of time; and the square of the orbital period of a planet is directly proportional to the cube of the semi-major axis of its orbit.

Figure 2.3 *The heliocentric or Sun-centered model places the Sun (in yellow) at the center of the universe with all the planets orbiting it in perfect circles.*

[40] Aiton, E. 1977. "Johannes Kepler and the 'Mysterium Cosmographicum.'" Accessed April 12, 2015. https://www.jstor.org/stable/20776469?seq=1#page_scan_tab_contents

Other scientists were also trying to understand terrestrial mechanics in relation to gravity. For instance, the Italian physicist Giambattista Benedetti (1530–1590), and the Dutch physicist Simon Stevin (1548–1620), did some work on the falling rock experiment. They were adamant that the rate of fall is independent of how heavy the falling body is.

Renaissance: Galileo Galilei

In the late sixteenth and early seventeenth centuries, Galileo Galilei (1564–1642) started work on gravitational theory. Born in Pisa, Italy, he studied medicine at the University of Pisa at the request of his father. The latter, a mathematician, kept his son away from the study of mathematics as he did not want it to interfere with his medical studies. Luckily, it was only a matter of time before Galilei discovered the wonder of geometry, which aroused his interest in mathematics and began, despite his father disapproval, to follow in his footsteps. His father eventually realized that he had no choice but to let Galileo study mathematics and science.

One of the greatest inventions that was to revolutionize astronomy was the telescope. Although attributed to Galileo, it was actually Hans Lippershey (1570–1619) who invented the telescope in 1608. Ironically, it was Galileo who made extensive use of it to build a strong body of evidence in support of the Copernican model. The astronomical and scientific observations made by Galileo were a blow to the Aristotelian/Ptolemaic model. For instance, his observations showed that celestial bodies were in fact covered with craters and mountains, in contrast to Aristotle's view that all celestial objects were smooth and perfect; the Sun was not perfect but had spots. The discovery of the four moons of Jupiter, as well as the observation of the phases of Venus, were further blows.

Galileo also studied the motions of pendula and bodies rolling and sliding down inclines. He went on to formulate his law of *Falling Bodies*. One of Galileo's most important contributions was the alleged experiment[41]

---

[41] The so-called experiment performed by Galileo may have been known or performed by others before him. If we are to believe the account in Aristotle's physics, we learn

conducted from the leaning tower in Pisa, where he showed that "all objects fall at the same rate, regardless of their mass." Galileo dropped two spheres of equal size but different masses from the top of the tower. One sphere was made of lead and the other of soap. In the experiment, both spheres reached the ground at the same instant upon being released simultaneously from the tower. Did Galileo's experiment ever take place from the leaning tower of Pisa? It is difficult to confirm this although written evidence seems to suggest that it did indeed take place during his time. In *Two New Sciences* Galileo[42] says: "But I, Simplicio, who have made the test can assure you that a cannon ball weighing one or two hundred pounds, or even more, will not reach the ground by as much as a span ahead of a musket ball weighing only half a pound, provided both are dropped from a height of 200 cubits." Other records point out that Galileo never did the experiment, but that he had discovered and understood the law of falling bodies from previous experiments.

Galileo's contributions to physics and particularly gravity are significant, from his first observation (published in 1610) to the discoveries of the satellites of Jupiter, the phases of Venus, and sunspots, etc. However, despite his immense contributions, he was not popular amongst the leading authorities of the time. In 1633 he was arrested and tried because of his astronomical model of the universe and for lending support to the Copernican view of the universe. He recanted under immense pressure and torture. He was later allowed to return to Florence, where he spent the rest of his life in seclusion working on the principles of mechanics, in particular on observations of the Moon and on other applications of the pendulum for clockwork.

---

that, in the fifth century a Byzantine philosopher named Iohannes Philiponus recorded and performed a Galileo-style experiment. The uniqueness of the Galileo or Philiponus experiments is that they require that two fundamentally different quantities, inertia and passive gravitational mass, are always proportional to one another. This simply means that the two quantities are equivalent measures of a single physical property, the quantity of mass of an object. Ironically, it is this discovery that led to the origins of the phrase equivalence principle.

42 Galilei, G. 1632. *Dialogue Concerning Two New Sciences*. Accessed July 10, 2017. *http://galileoandeinstein.physics.virginia.edu/tns_draft/*

Enlightenment: Sir Isaac Newton

Further support for the heliocentric model came from the work of Newton (1643–1727). Following the publication of *Philosophiae Naturalis Principia Mathematica* in 1687, and based on the work of the Egyptians, Galileo, and Kepler, the heliocentric model was fully established. Newton argued that the force that makes planets move around the Sun is the same force that makes object fall on Earth: that force is called gravity. According to Newton, gravity is a force of mutual interaction of bodies with mass, and this force is inversely proportional to the square of the distance between the objects.

With the force that pulls everything together fully explained, Newton finally established the heliocentric model. Some unanswered questions remained however, such as action at a distance; as Newton himself was not comfortable with the notion of gravity, he never "assigned the cause of this power." This is because he was not able to theoretically identify or experiment on the motion that produces the force of gravity. However, despite his reservations about the origin of gravity, Newton reportedly told four people that his theory of gravity was inspired by an apple falling from a tree. This story continues to be hotly debated and disputed.

Having said that, Newton's immense contributions to gravity are perhaps more important than the work of his predecessors; he was the first person to put together and explain the laws that sum up all the previous discoveries of the Egyptians, Muslims, Greeks, Romans, Kepler, and Galileo. Secondly, he concluded that the laws that apply here on Earth are the same that apply to the movement of celestial bodies. Following Robert Hooke's suggestion that there is a gravitational force between bodies which depends on the inverse square of the distance between them, Newton was able to derive Kepler's laws in the late seventeenth century. Moreover, in 1666 he wrote[43]:

> *"I deduced that the forces which keep the planets in their orbs must be reciprocally as the squares of their distances from the centers about*

---

[43] Simmons, G. S. 2007. *Calculus Gems: Brief Lives and Memorable Mathematics*. The Mathematical Association of America: USA.

> *which they revolve, and thereby compared the force requisite to keep the moon in her orb with the force of gravity at the surface of the earth and found them to answer pretty nearly."*
>
> —SIMMONS 2007, 139

Thirdly, he summed up the laws that could explain the behavior of earthly and celestial bodies. Above all, one of the great successes of Newton's work was its accurate prediction of the existence of Neptune based on the motion of Uranus; a discrepancy in Mercury's orbit pointed out flaws in Newton's theory. Finally, in 1686 he stated that "all sorts of heavy bodies (allowance being made for the inequality of retardation which they suffer from a small power of resistance in the air) descend to Earth from equal heights in equal times".

Twentieth Century: Albert Einstein

Newton's theory of gravity was still dominant when Albert Einstein (1879–1955) was born, despite some discrepancies in the theory that could not be explained, namely a gap in Mercury's orbit. Experimental results by astronomers showed that the orbit of Mercury around the Sun is not a fixed ellipse. This was not predicted by Newton's theory since all eclipses should be fixed. In 1915 Einstein postulated his new theory of gravitation in the form of the general theory of relativity (GR), in which gravity is not a force but simply a consequence of the curvature of space-time. From GR, we can now understand why a massive object creates a curve in space-time and the inertial trajectory or geodesics that cause straight lines to become curved.

In1916 he managed to find a theory that would explain the flaw in Newton's theory. It accounted for the small discrepancy in Mercury's orbit. He later extended the Equivalence Principle and incorporated it as a postulate for his theory of GR. The Equivalence Principle states that "all of the laws of physics are the same in all small regions of space, regardless of their relative motion or acceleration."Thus, the theory of GR rests on a single and unique experimental fact that all objects fall with the same acceleration. In Einstein's GR, the effects of gravitation are due to space-time curvature rather than a force.

The motion does not cause space-time curvature, but space-time curvature is the cause of the motion. Many years passed after the discovery of GR before it was realized that it was not a complete theory of gravity due to its incompatibility with QM. attempts to describe gravity in the framework of quantum field theory, despite some successes, also fail at short distances on the order of the Planck length. I wonder what went on in Einstein's mind when it was discovered that the universe was expanding. The limitations of his theory raised important questions as we learned from gravitational theory that all the matter in the universe was supposedly attracting each other[44].

Einstein was not satisfied with his achievement. He continued for at least thirty years to formulate a theory that could solve these problems and unify physics. I have been asking myself some fundamental questions about what Einstein knew or not about gravity:

- Why did Einstein never debunk the myth of gravity?
- Did he know that something in GR was fundamentally wrong?
- Did he discard any of his findings? Nobody is certain, because some of his statements suggest that he was not particularly happy with his theory of GR and had some reservations.

A quote by Pais[45] in 1982 suggests so:

> *I [Einstein] consider it quite possible that physics cannot be based on the field concept, i.e., on continuous structures. In that case nothing remains of my entire castle in the air gravitation theory included, [and of] the rest of modern physics.*

---

44 It is the hypothesis that DM and dark energy comprise 96% of the missing matter of the universe, thus forcing it to expand. However doubts continue to be raised by many physicists. GR is so far poorly supported.

45 Pais, A. 1982. *Subtle is the Lord*. ISBN-10: 0192851381, Oxford University Press: New York.

As far as 1916, he led a call for a serious second look at the theory of gravitation:

> *"...due to the inner-atomic movements of electrons, atoms would have to radiate not only electromagnetic, but also gravitational energy, if only in a tiny amount. As this is hardly true in nature, it seems as if QT will have to modify not only Maxwellian electrodynamics, but also the new theory of gravitation."*
>
> —EINSTEIN 1916B, P. 696

Future Directions

The current level of technological and engineering development achieved by humanity has taken centuries of hard work by scientists and engineers. However, so far, the technology in our hands has not reached the stage where we can develop techniques to escape gravity without using sources such as rockets. How do birds and flying insects escape gravity? It begs the question: how could ancient Egyptians have developed technology to escape gravity without using huge force? How did they build huge pyramids without mastering gravity?

Is gravity simply a form of attractive force, and not a separate force as we know it? What is the origin of gravity? Who knows what the future will bring—isn't science an ongoing process? In what follows, I would like to highlight some anomalies and discrepancies in the theory of gravitation as well as possible solutions. Perhaps the key to understanding and unlocking the secret of gravity is to find its origin. This is of paramount importance. The history of gravitational theories is summarized in Table 2.1

Table 2.1 *History of gravitational theory from Egyptian times to the twenty-first century.*

Timeline	Contributions
Ancient Egyptians	• Scientific contributions to science, mathematics, astronomy, architecture, physics, chemistry, etc. • The universe resembles a large rectangular box with Egypt at the center. • Had discovered the movements of the planets around the Sun. • Worshiped the Sun because of its central position in our solar system. • After studying the Egyptian system, Sir Isaac Newton was convinced that they knew about gravity and were acquainted with atomic theory.
Ancient Greeks	**Aristotle: Geocentric Model** • Earth is at the center of the universe. • The Sun and the Moon move in a circle around the Earth. • Planets circle the Earth and all objects fall onto it. • They believed in the existence of four natural elements (earth, water, fire, and air). • All objects fall towards Earth. **Aristarchus: Heliocentric Model** • The Sun was the center of the universe and the Earth revolved around it. • The Sun was much larger than the Earth, therefore the small body should orbit the larger one.
Middles Ages	**Nicolaus Copernicus: Heliocentric Model** • The Sun is the center of the universe. • Earth and other planets orbiting the Sun. • Model could not account for why the planets circled the Sun. **Tycho Brahe** • Discovered a new star at the age of 26. • Earth is at the center of the universe. • The Sun moves around the Earth, and the rest of the planets circle the Sun

Timeline	Contributions
	Giambattista Benedetti (1530–1590) **and Simon Stevin** (1548–1620) • Did some work on the falling rock experiment, and were adamant that the rate of fall is independent of how heavy a body is.
Renaissance: Galileo Galilei	**Galileo Galilei** (1564–1642) • Significant work on gravitational theory. • Studied the motion of pendula and bodies rolling and sliding down an incline. • Formulated the Law of Falling Bodies. • His astronomical observations were a blow to the geocentric model. • Discovered the four moons of Jupiter. • Observed the phases of the Moon. • The Sun was not perfect; it had spots.
Enlightenment: Sir Isaac Newton	**Isaac Newton** (1643–1727) • Immense contributions to understanding gravity. • Summed up the laws that explain behavior of earthly and celestial bodies. • Accurate prediction of the existence of Neptune based on the motion of Uranus. • Uncomfortable with the notion of gravity as he never assigned the cause of this power. • Inverse square law (action at a distance). • Newton established the heliocentric model.
Twentieth Century: Albert Einstein	**Albert Einstein** (1879–1955) • The Principle of Equivalence: all bodies are accelerated identically by a gravitational field. • Gravity and space-time (light gets bent by acceleration or by gravity). • Curved geometries: distributions of mass and energy will bend space. • Predictions from General Relativity (GR).

Timeline	Contributions
	Bending of starlight (observed during total eclipses of the Sun (1912, 1914, and 1919). • Mercury's perihelion (Newton: advance of 5514 arcsec/century; observations: 5557 arcsec/century). • Gravitational redshift (electromagnetic radiation is redshifted by strong gravitational fields).
Twenty-First Century	• What will the future bring? • Newton never attempted to offer an explanation for what gravity might be. • Origin of gravity? • Gravity as information? • Gravity as an electromagnetic force? • Sakharov's induced gravity? • MOdified Newtonian Dynamics (MOND)?

Anomalies and Discrepancies

The theory of gravity developed by Newton and later extended by Einstein has been successful and is one of the most remarkable triumphs of scientific achievement. Observation vindicates numerous predictions of this theory with accuracy. For instance, the successes of Newton's theory of gravity abound, and there is hardly any need to point this out. In the same way, Einstein gravity is even more successful because many empirical observations match the theoretical predictions. However, with almost all theories there are cases where it is no longer applicable. I would like to show how other theories of gravity, some not as successful as Einstein's theory and other perhaps more elegant, are offering alternatives to Einstein's theory. Einstein's theory is being challenged by new theoretical alternatives plus the accumulated observational evidence that suggests that some phenomena are not sufficiently accounted for or cannot be explained by our current theory—this may suggest the need for a new theory of gravity.

Some possible gravitational anomalies and discrepancies have been pointed as a result of experiments and observations over many years. So far there are no satisfactory conventional explanations for these phenomena, although various papers have offered plausible explanations. Among the possible gravitational anomalies observed are the Allais effect that occurs during eclipses and the Pioneer anomaly; these were investigated by the Pioneer 10 and 11 and Ulysses spacecraft. These anomalies hint at possible departures from the known gravitational laws. They point to the need for new physics to account for these behaviors. Perhaps the entire foundation of cosmology based on Einstein's equations must be revised, simply because so far BHs, the big bang, gravitational waves, dark energy, or DM have never been observed. Above all, the notion of DM and dark energy, which were invented to justify the theory of gravity, have so far remained elusive or not matched tangible observations; they may have to be abandoned.

Some of these unaccounted for anomalies, discrepancies, and problems with current theories of gravity indicate that the SM of cosmology is in deep trouble:

- There is irrefutable evidence that some stars seem to be older than the age of the universe, thus questioning the validity of the big bang model.
- The problem of the missing DM which so far has eluded observation.
- The universe seems to be expanding faster over time rather than more slowly.
- The Allais effect: this effect was discovered by the respected French physicist Maurice Allais on June 30, 1954. He observed that a Foucault pendulum exhibited anomalous behavior (movements) during the time of a solar eclipse. Natural explanations based on conventional effects such as seismic activity, changing temperature, atmospheric pressure, magnetic, lunar or solar effects, were ruled out.
- The Pioneer anomaly: or the anomalous curves of galaxies, was first observed in 1980 by John Anderson.

- The Flyby anomaly: this is characterized by an unexpected increase in acceleration of some spacecraft during slingshot maneuvers. This unexpected energy increase during Earth flybys causes an unaccounted-for velocity increase of over 13 mm/s.
- The missing neutrinos from the Sun which have so far remain unaccounted for.
- The speed of light seems to be getting slower over time.
- There is some evidence to suggest that the universe may be shaped like a donut; data analysis from WMAP suggests this, contradicting current theories about the big bang hypothesis.
- Some experiments cannot be understood in the frame of GR, for example, the impulse gravity generator.

ALTERNATIVE THEORIES

The fundamental nature of gravity is still unknown and not well understood. Attempts to unify gravity with three other forces, or to develop a quantum theory of gravity (QTG), have proven difficult—and one of the main reasons might be the lack of understanding of the origin of gravity. It is because of this that some physicists, mainly theorists, have been leading efforts to understand and unify gravity with other forces. Recent advances in supergravity or ST are illustrative of the constant struggles that physicists are facing to reach this goal. In the sixties a renowned Russian physicist, Andrei Sakharov, put forward a radical hypothesis that gravitation might not be fundamental at all, but a residual effect associated with other fields. He designed a model in which gravity could be viewed as an elasticity of the space-time medium.

Since then Sakharov's work has been extended by some physicists and the results look promising. Recent advances and developments are well illustrated in the work of Puthoff, and can be used as a springboard to understanding the origin of gravity. The model proposed by Sakharov might be useful for providing hints on how gravity emerges. The failure of Newtonian dynamics and GR beyond Earth is that these two theories cannot explain many things, for example, observed galactic dynamics and cosmological systems like galaxies, small groups of galaxies, binary galaxies, and above all rich galaxy clusters. These two theories are rescued from collapse only if matter-energy ingredients are introduced into the universe in the form of DM and dark energy. One of the most important alternatives to reject DM is MOdified Newtonian Dynamics (MOND), the brain-child of Mordehai Milgrom, an Israeli physicist. In 1983 he proposed a specific change to the equations of Newton which govern particle motion at very low accelerations.

Recently, Verlinde suggested that the key to unlocking gravity is information. Before him, in 1995, Ted Jacobsongeneral relativity" showed that Einstein's equations of GR could be derived using thermodynamics and the HP, thus hinting at the important role of information. Both works underline the desire to look again at gravity as we know it, and suggest that we should be ready to abandon the notion of gravity as a fundamental force. Some alternatives have been proposed to address the problems identified in the standard cosmological model (anomalies and discrepancies) as well as the issue of the origin of gravity.

The outcome of this ongoing dialogue will have significant implications for any ToE and particularly for QG. It is also possible to argue that several current attempts to quantize gravity may end up failing because some schools of thought have suggested that there are no fundamental gravitational degrees of freedom to quantize. Some have even argued that Einstein's gravity is an emergent phenomenon, in the same way that fluid dynamics emerges from molecular physics as a low momentum long-distance approximation.

The history, aims, results, and future prospects of some of the proposed alternatives are now discussed. The discussion starts by presenting and exploring the rich history of induced gravity, as the last decade has seen remarkable and sustained development of induced gravity models. Furthermore, the contribution of Milgrom is looked at in detail, as is the latest suggestion by Verlinde. It is now a common opinion (perhaps among a minority) that gravity and the notion of space-time and space-time geometry might not be fundamental at all.

Andrei Sakharov's Induced or Emergent Gravity

Perhaps one of the most prominent of these alternative approaches is Sakharov's induced gravity. Sakharov[46] argues that gravity is a force induced by residual electromagnetic forces rather than a fundamental physical field. It emerges from quantum field theory in the same way, for instance, that hydrodynamics emerges from molecular physics. Sakharov originally proposed induced or emergent gravity in 1967. He was inspired by many condensed matter systems that give rise to emergent phenomena which are identical to GR. In the same year Zel'dovich published a parallel result concerning electrodynamics.

Recently, many authors have argued that models of emergent gravity such as the one proposed by Sakharov and others have been proven impossible by the Weinberg-Witten theorem and the work of Jake Todd[47]. However, many works have shown that models with emergent gravity are possible if space-time dimensions emerge together with gravity; a typical example is the Maldacena AdS/CFT correspondence in ST.

What is induced gravity? The proponents of this school of thought have argued that gravity is not fundamental, but is induced by quantum fluctuations of the vacuum. Induced gravity emerges from zero-point energy (ZPE). As such, ZPE and gravity are made up of massless particles, and ZPE is spread out in space-time; it is this ZPE that causes objects to move towards a body that has caused space-time to stretch because of the pressure differences as one moves from one area of stretched space to another.

Using the same line of reasoning, Sakharov suggested that gravity might be an induced effect due to changes in the ZPE of the vacuum because of the presence of matter. The work of Sakharov has been developed and extended

---

46 Sakharov, A. 1967. "Vacuum Fluctuations in Curved Space and the Theory of Gravitation." Translated from *Doklady Akademii Nauk SSSR* 177 (1):70–71.

47 Todd, J. 2010. "Disproving Induced Gravity and Induced Inertia Theories Related to Zero-Point Energy." Accessed June 16, 2016. http://www.gsjournal.net/Science-Journals/Research%20Papers-Relativity%20Theory/Download/3910

by many researchers. Puthoff[48], particularly, has addressed in detail many aspects of this approach and made some outstanding contributions. He has also been one of the leading advocates of induced gravity and has done a great service to physics—his contributions are unique, profound, and deeply enriching. His approach is very simple: ZPE causes matter to move around quickly and, as a result, it causes electromagnetic forces to exist which give rise to gravity.

This quote summarizes his reasoning:

> *"The gravitational interaction is shown, to begin with the fact that a particle situated in the sea of electromagnetic zero-point fluctuations develops a "jitter" motion, or ZITTERBEWEGUNG as it is called. When there are two or more particles they are each influenced not only by the fluctuating background field, but also by the fields generated by the other particles, all similarly undergoing ZITTERBEWEGUNG motion, and the inter-particle coupling due to these fields' results in the attractive gravitational force."*

Because of the gravity induced by electromagnetic forces approach, some theorists have gone as far as to suggest that Sakharov's induced hypothesis is already a unified theory because it provides a basis for understanding many of the various unexplained characteristics of the theory of gravitational interaction. An excellent account of the successes of induced gravity is illustrated in an interesting article entitled "Induced Gravity and Gauge Interactions Revisited," by Boguslaw Broda and Michal Szanecki[49], first published in 2009.

---

48 Puthoff, H.1989. "Quantum Fluctuations of Empty Space: A New Rosetta Stone in Physics?" Accessed January 12, 2016. http://www.p-i a.com/Magazine/Ref/Puthoff_Rosetta_ZPE.html

49 Broda, B. and Szanecki, M. 2009. "Induced Gravity and Gauge Interactions Revisited." Accessed July 13, 2010. http://arXiv:0809.4203v4 [hep-th]

Zero-Point Energy

The origin of gravity as proposed by Sakharov is embodied in the concept of ZPE. One of the predictions of QT (which has been experimentally verified) is that space or the vacuum contains residual background energy, called ZPE. QT predicts that all the space must be filled with electromagnetic zero-point fluctuations or a zero-point field, thus creating a huge, universal sea of ZPE. Why ZPE? Experiments have demonstrated beyond any doubt that at the absolute zero of temperature (-273°C), elementary particles continue to exhibit energetic behavior.

Looking at induced gravity in terms of ZPE, one begins to understand why some theoretical physicists have suggested that ZPE is the key to understanding inertia and gravity, and why it perhaps controls these forces. Some also see the possible existence of ZPE as an opportunity to tap energy and use it for our benefit. Moray King, for instance, has argued that there is evidence to show that ZPE is not a passive system but is a manifestation of an energy flux passing through our space orthogonally from higher dimensions. The extension of this concept is found in the work of Wheeler and is well illustrated with the derivation of hyperspace channels or wormholes.

It is possible that some of the most important conundrums in physics, particularly the non-local connection EPR experiment and Bell's theorem, could be understood from the ZPE perspective. Where does the ZPE come from? An interesting point about the origin of ZPE in quantum physics comes from the Heisenberg uncertainty principle. The uncertainty principle relates the uncertainty in the position and momentum in the measurement of a quantum particle in QM. Heisenberg used the term (ΔQ) to indicate the uncertainty in the position measurements around some average value, and (ΔP) for the spread or uncertainty in the momentum measurements. Here we refer to some average value in a series of measurements. We learn in QT that, regardless of the quantum device, the products of the uncertainties can never be less than Planck's constant, $\hbar$. For instance, if your quantum measuring device is excellent at determining the position term (ΔQ), you will lose information about the momentum (ΔP).

The notion of uncertainty is widely found in many other areas of physics, mathematics, and information theory, etc. For instance, there is a similar uncertainty relation between the energy of a particle (ΔE) and the elapsed time (Δt). The energy-time uncertainty principle was well known by the founders of QM: one cannot determine the exact energy and the elapsed time simultaneously. Given a series of measurements, the product of the uncertainty of the energy and the uncertainty of the elapsed time is always greater than or equal to $\hbar$. Likewise, in information or communication theory, one can exactly measure either the frequency or the time of occurrence, but not both at the same time. One can measure the frequency with arbitrary precision, but there is a limit to the precision when these two measurements are taken simultaneously: one cannot simply measure both frequency and time precisely, at the same time.

This means that if the frequency information is exactly known, one would not know anything about the time it occurred. Similarly, in brain research we learn that when frequency and temporal measurements are made simultaneously, there is a limit to the precision of the measurements which is determined by the uncertainty principle. According to Pribham, the brain functions as a dissipative structure to seek to decrease this uncertainty in the direction of its theoretical limit. This uncertainty is not due to any flaws in measurement, but simply reflects an intrinsic quantum fuzziness in the nature of energy and matter, which arises from the wave nature of the various quantum fields; this leads to the concept of ZPE. This ZPE is defined as the energy that remains when all other energy is removed from a system.

Various examples in nature show ZPE in action. For instance, consider liquid helium at a certain temperature. If we decrease its temperature to absolute zero (-273°C) the experiment shows that helium never freezes to a solid but remains a liquid, due simply to the irremovable ZPE of its atomic motion. Consider the quantum harmonic oscillator; its motion cannot be brought to rest, unlike the classical harmonic oscillator with mass on a spring that can easily be brought to rest. For the quantum oscillator, a residual motion will remain due to the requirements of the Heisenberg uncertainty principle.

The Casimir force is constantly referred to as evidence in support of the Sakharov hypothesis: there must be a sea of ZPE underlying the universe. What is this Casimir effect? In 1947 the work of Hendrik Casimir (and the Dutch physicist Overbeek) on the theory of Van der Waals forces led to the discovery of what is known today as the Casimir force. According to Casimir[50], Bohr "mumbled something about zero-point energy" being relevant during their discussion. This suggestion by Bohr led Casimir to look closely at the concept of ZPE effects in the related problem of forces between perfectly conducting parallel plates[51].

Scientists have known for some time that a vacuum is not empty, but is indeed filled with particles and energy. The presence of these quantum particles can produce energy because the vacuum is not empty. ZPE has rekindled much interest in the study of the vacuum as its existence has been proven experimentally via the detection of the Casimir effect. The mid-1990s saw an increased number of physicists looking at it very carefully. For instance, Steve Lamoreux[52] of the University of Washington verified Casimir's theory to within five percent in the size range of a few microns. Umar Mohideen also verified Casimir's predictions more precisely than his predecessor. Since then other researchers have successfully repeated and refined the experiment.

Proof that ZPE exists has led to some investigators attempting to find ways of tapping the ZPE and extract it for use in other applications. In 1993, a

---

50 Maclay, G., Fearn, H. and Milonni, P. W. 2001. "Of Some Theoretical Significance: Implications of Casimir Effects." Accessed March11, 2017. https://arxiv.org/pdf/quant-ph/0105002.pdf

51 We learn that the cavity between conducting parallel plates cannot withstand all modes of the electromagnetic field. It had been argued that there is a zero-point radiation outside the plates which pushes them together—the resulting effect is now called the Casimir force. The force increases in strength with the inverse fourth power of the plate separation, and becomes null (ceases) when the plates come into contact with each other (the Casimir force acts to push the plates together). The Casimir effect is a force produced solely by activity in the vacuum. In addition, the Casimir force is very powerful at small distances.

52 Lamoreaux, S. K. 2000. "Experimental Verifications of the Casimir Attractive Force Between Solid Bodies." *Comm. Mod. Phys. D: At. Mol. Phys.* (2):247–61.

paper by Cole and Puthoff[53] entitled "Extracting Energy and Heat from the Vacuum," suggested that it is possible in principle to extract energy and heat in a vacuum without violating the laws of thermodynamics. This development led Robert Forward to design a thought experiment in 1984 on how to build a "vacuum fluctuation battery." Eugene Mallove and Bernard Haisch have also spent many years trying to find ways to harness and use this free energy that permeates the universe.

We also learn that the existence of the ZPE was inferred much earlier by Einstein, Planck, Nernst, and other researchers in the context of blackbody radiation. Furthermore, Einstein and Otto Stern were close to deriving the blackbody function without the need of quantization but with the presence of ZPE. As far back as 1916, Nernst claimed that the universe was filled with ZPE. In the 1930s even the great English physicist Paul Dirac argued that the vacuum was filled with particles in negative energy states. Unfortunately, with the advent of QT, this line of inquiry was abandoned.

In direct relation to gravity, Newton's work, and the ZPE, Bernard Haisch[54] of Calphysics suggested that:

> *"There exists a background sea of quantum light filling the universe, and that light generates a force that opposes acceleration when you push on any material object," [argues Haisch]. "That is why matter seems to be solid, stable stuff that we, and the world, are made of. So maybe matter resists acceleration not because it possesses some innate thing called mass as Newton proposed and we all believed, but because the zero-point field exerts a force whenever acceleration takes place."*
>
> —HAISCH 2001

53 Cole, C. D. and Puthoff, H. E. 1993. "Extracting Energy and Heat from the Vacuum." *Physical Review* A. 48(2):1562–1565.

54 Haisch, B. 2001. "Brilliant Disguise: Light, Matter and the Zero-Point Field." Accessed February 05, 2015. http://altered-states.net/barry/update269/zeropointfield.htm

The work of Puthoff[55] on inertia and gravity follows the same reasoning as Haisch. He showed that inertia and gravity are proof of ZPE by deriving Newton's law (force is equal to mass times acceleration) from ZPE electrodynamics and demonstrated that it might be related to the known distortion of the zero-point spectrum in an accelerated reference frame. According to Puthoff, the resistance to acceleration defines the inertia of matter; it appears to be an electromagnetic resistance. Puthoff's work suggests that the inertia effect is a distortion of the ZPE field at high frequencies, in contrast to the effect of gravity which has been shown to be a low-frequency effect.

To sum up, I would like to stress that the experimental evidence for the existence of ZPE includes the Casimir effect, the Lamb shift, Van der Waal's force, diamagnetism, spontaneous emission, very low temperature liquid helium, etc. So far, there is no known alternative explanation that accounts for the source of the energy which keeps helium from freezing, apart from ZPE. The ZPE is the energy left behind in a volume of space when all the matter and radiation has been removed. ZPE, also known as vacuum fluctuation energy, gives rise to real measurable phenomena such as the Lamb shift and the Casimir effect.

ZPE is associated with all natural force fields, including the electromagnetic field. These fields associated with ZPE produce physical effects that are well documented. For example, the Lamb shifts in the spectral lines of an atom. In the Lamb shift the fields slightly perturb an electron in an atom so that, when it makes a transition from one state to another, it emits a photon whose frequency is shifted slightly from its normal value. From Puthoff's work and other related research on ZPE, it appears that an exchange of energy ultimately connects everything in the universe. Although the last word has not been uttered on whether or not the Casimir effect is evidence of ZPE, some researchers such as Christian Beck and Michael Mackey have gone so far as to suggest that dark energy is nothing more than ZPE.

---

[55] Puthoff, H. 1994. "Inertia as a Zero-Point Lorentz Force." *Physical Review* A 49 (2):678–694.

The Problem with Dark Matter Theory

In the 1970s Vera Rubin[56] and colleagues used the Doppler shifts of spectral lines to investigate galaxy rotations—to establish rotation curves for many spiral galaxies. Contrary to their expectations based on earlier work by astronomers, the rotation curves in the outer regions of galaxies stayed roughly the same, or in some cases rose slightly toward the outer edges of the galaxies. The expected outcome was that the orbital velocities outside the nuclear bulge would decrease with distance from the center in a Keplerian fashion.

The result of this investigation meant that the existing Newton's law of gravity failed to explain the motion of galaxy rotation. This had far-reaching consequences and is known today as the mass discrepancy problem: the discrepancy between galaxy masses inferred dynamically and from their emitted light distributions. This result suggested that the galaxies were accompanied by a vast amount of undetected matter, known as DM. This led to the DM hypothesis, which was later incorporated into the standard cosmological model.

The incorporation of DM in the standard cosmological model has become one of the most famous puzzles in cosmology. Newton's theory of gravitation fails to explain the motions of stars and gas in galaxies as well as in larger systems. This realization has led to the inference that most of the mass in the universe (approximately 95%) is dark. The need for DM arises in the context of standard cosmology, as the universe has been expanding since the initial big bang and, accordingly, the early universe contained a uniform hot mixture of matter made from electrons, protons, photons, and DM.

Dwarf stars (white and brown), BHs, and neutrinos—as well as hypothetical objects such as gravitinos and magnetic monopoles—have been proposed to account for the DM. In fact, we should be talking about mass discrepancies rather than looking for evidence of DM. Is the universe full of DM or is it simply

---

[56] Rubin, V.C., Ford, W.K., Jr. and Thonnard, N. 1980. "Rotational Properties of 21 Sc Galaxies with a Large Range of Luminosities and Radii, from NCG 4605 (R = 4 kpc) to UGC 2885 (R = 122 kpc)." *Astrophysical Journal* 238:471–487.

that the theory behind it (which leads to the inference of DM) needs to be reviewed? It is a fact that Newton's universal law of gravity has been successful, but it may not be accurate in galaxies or other extragalactic systems. Standard cosmology proposes a flat inflationary big bang universe. In this model, the mass density of the universe consists of dark energy and DM, as well as visible matter such as stars, planets, gas, etc. Accordingly, the formation of galaxies is attributed to density variations in the distribution of DM in the supposed early universe.

The hypothesis of DM has been called into question many times in the astronomical and cosmological literature. This has led some cosmologists and astronomers to propose a modification to our law of gravity. Many such attempts have been put forward but have failed to stand the test of time. Over the years alternative interpretations of the mass discrepancy have been presented. Unfortunately, most of them have been rejected, sometimes without careful consideration and investigation. As early as 1923, James Jeans[57], one of the leading astronomers in England, pointed out the need to modify Newton's law of gravity particularly as it related to gravity on galactic scales.

In 1963, Arrigo Finzi[58], in an article entitled "On the Validity of Newton's Law at a Long Distance," proposed a distance-based modification of gravity to deal with the mass discrepancy in galaxy clusters as a possible alternative to the DM theory. However, his work was shown to be inconsistent by Milgrom in the 1980s, who argued that such modifications of the distance of gravity fail to reproduce the observations. Similarly, in 1990, work by Wright, Disney, and Thompson[59] entitled "In Universal Gravity: Was Newton, Right?" suggested that a modified (inverse linear) law of gravity beyond a certain distance scale may explain the mass discrepancies in galaxies and galaxy clusters.

The problem is that there is not yet any known experimental evidence for DM. We are aware that there are two types of particles that are the leading

---

57 Jeans, J. H. Sir. 1929. *Astronomy and Cosmogony*, 2nd Ed. Cambridge University Press: Cambridge.

58 Finzi, A. 1963. "On the Validity of Newton's Law at a Long Distance." Monthly Notices of the Royal Astronomical Society 127:21–30.

59 Wright, A. E., Disney, M. J. and Thomson, R. C. 1990. "Universal Gravity: was Newton Right?" *Proceeding of the Acoustical Society of America* 8 (4):334–338.

candidates for cold dark matter (CDM). The two most prominent of these are supersymmetric extensions of ordinary particles and the axion. We note that CDM such as neutrinos, massive dead objects, and stellar objects have so far been discarded. The two types of CDM mentioned here are hypothetical; neither of them has been produced in the laboratory or observed anywhere. A cautionary approach to the idea of DM should be adopted. Perhaps what we think of as evidence for DM may simply be the failure or breakdown of Newton's and Einstein's gravity and dynamics?

The problem with Newtonian dynamics and GR is that these two theories fail to account for such things as the observed dynamics of most galactic and cosmological systems like galaxies, small groups of galaxies, binary galaxies, and above all rich galaxy clusters. These two theories can only be successful if mater-energy ingredients are introduced into the universe in the form of DM and dark energy. One of the proponents of an alternative to DM, Stacy McGaugh[60], argued in 2002 that:

> *"But what if, instead of extra mass in the form of dark matter, there was extra gravity? In other words, over the scale of a galaxy, gravity pulled a little harder than we expect based upon our knowledge of its behaviour in and around Earth. What if we could modify our laws of gravity so that they provide that extra force at long range, while leaving their behavior on smaller scales untouched? If the idea holds, we would have no need for dark matter."*
>
> —MCGAUGH 2002, 64

Earlier works by astronomers and physicists have pointed out this discrepancy. It seems likely that Newton may have indeed been wrong, and that his gravity theory is not universal after all. His theory does describe the everyday effects of gravity on Earth and in the solar system, however, it is possible that it is not applicable and fails in other spheres such as galaxies, clusters, etc. Perhaps we have so far failed to comprehend the physics underlying the

---

60 McGaugh, S. 2002. "Mond over Matter." *Astronomy Now* (January):63–65.

force of gravity. In the next section, I introduce two of the most important alternative models to date, Milgrom's MOND and that of emergent gravity as proposed by the Dutch theoretical physicist Erik Verlinde.

Mordehai Milgrom's MOdified Newtonian Dynamics (MOND)

One of the most important alternative models to date is MOND, the brainchild of Milgrom, an Israeli physicist; in 1983 he proposed a specific change to the Newton equations governing particle motion at very low accelerations. The theory has since been extended by Bekenstein who has published a relativistic version of MOND known as tensor-vector-scalar theory, an extension of GR. One of the most pressing problems in cosmology is how to explain rotation curves at large galactic radii. DM is introduced by the SM to explain the observational evidence of flat rotation curves in disk galaxies.

Milgrom suggests a possible explanation for the flat rotation curve through a modification of Newton's second laws without the need to invoke DM. According to MOND, galaxies do not have a significant dark matter halo. One of the supporters of MOND, Stacy McGaugh, argues that the proposed theory reduces to the usual Newtonian form in the regime of high acceleration, but at accelerations lower than a hundred-billionth of what we feel here on Earth things change in a way that might account for the mass discrepancies in galaxies. What MOND theory is trying to achieve is very elegant and exciting. It incorporates standard dynamics at high accelerations, while forgoing them at low accelerations to explain all aspects of the mass discrepancies in the galactic system without the need for DM.

MOND theory introduces into physics a new constant a_0. The role of this constant is like that of the speed of the light c, in the GR context (departure from classical physics). In this context, in systems with acceleration larger than a_0 ($a>a_0$), the theory reduces to the standard Newtonian theory. This is simply achieved by substituting everywhere the MOND equation $a_0=0$. The theory proposes that there is an acceleration a_0 under which gravity is stronger than is predicted by Newton's theory. When the acceleration is much larger than a_0 Newton's second law remains intact. However, when the acceleration is smaller than a_0, Newton's second law is altered and the force becomes proportional to the square of the acceleration.

MOND considers the observed acceleration in galaxies to be indicative of a smaller force. Thus, we can do away with the need for DM. The theory suggests that the acceleration decreases with distance in the outskirts of galaxies, eventually going below the critical acceleration. The theory applies at very low accelerations (outer regions of spiral galaxies, galaxy groups...). The critical acceleration below which MOND applies is $a_0 = 1.2 \times 10^{-10} ms^{-2}$. This value is about one hundred billion times smaller than the acceleration due to gravity on the surface of the Earth. By using MOND with a universal value of a_0, the rotation curves of many galaxies are reproduced without the need to invoke DM.

MOND Successes, Achievements, and Challenges

- The theory has been successful in explaining the motion of stars around the centers of galaxies. For instance, the result of rotation curves calculated using MOND have been compared with those observed in over one hundred galaxies, and so far no case has been found where MOND fails. Rotation curves for low brightness galaxies which have been available for quite some time also provide a test of MOND; so far, the results are good.
- A comparison between MOND and DM predictions for a sample of a spiral galaxies with accurately measured rotation curves shows beyond any doubt that MOND fits the rotation curves and provides the best description.
- Above all, MOND has been successful at predicting new phenomena—this is well illustrated in the original paper by Milgrom in 1983 where he made some predictions relating to then-unknown classes of galaxies.
- Astronomers have since discovered these galaxies, known as low-surface-brightness galaxies. Astronomical observations back MOND's predictions for the rotation curves of galaxies. So far there is no known astronomical evidence for DM.
- MOND obviates the need for DM in galactic systems. It predicts and explains an extensive body of galactic data.

- The opponents of this theory suggest that undetected DM dominates clusters, although so far there is no evidence for this; it has only been inferred. What is known is a mass discrepancy in galaxies in other galactic systems.

MOND Weaknesses, Failures, and Challenges

- Opponents of MOND argue that the theory has been unable to explain the motion of galaxies in rich clusters. In 1984, Felten[61] pointed out that the original formulation of MOND violates a basic principle underlying physics, namely that Newton's third law and linear momentum are not conserved. However, it appears that the shortcomings pointed out have been addressed and explained by Milgrom and Bekenstein.
- Many have argued that the theory fails to explain the mass discrepancy everywhere and if it is to be taken seriously, it must be able to explain the velocity dispersions of elliptical galaxies, the motions of galaxies in the large-scale universe, as well as the gas temperature of cluster galaxies.
- The theory fails convincingly to explain the need for DM in the cosmological setting and in galaxy clusters.
- Although MOND reduces substantially the amount of DM required, opponents of the theory have pointed out that there is still a remaining discrepancy between the required and observed masses in clusters. However, proponents of the theory think that this might simply be due to normal matter that has remained hidden in clusters.
- The difficulties of the SM abound when applied to galaxies, so that over the years, we have seen high numbers of astronomers convert to MOND. MOND must be formulated in a way that can go beyond the Newtonian theory of gravity, and provide a concrete solution to gravitational lensing and cosmic expansion.
- MOND predictions about the temperature of hot gas in clusters of galaxies disagree with observations.

---

61 Felten, J. E. 1984. "Milgrom's Revision of Newton's Laws—Dynamical and Cosmological Consequences." *Astrophysical Journal*, Part 1 (ISSN 0004-637X) 286:3–6.

To sum up, MOND is a new approach to understanding gravity beyond Newton and Einstein. It suggests that it may be possible to do away with DM. The theory points out that the problem does not lie with the amount of matter present (or DM) but with the laws of gravity governing its motion. So far, both Newton's law of gravity and GR have withstood many tests. Because of the successes, the overwhelming majority of the physics community believe that there is no need to discuss it and that there is nothing new to learn about gravity. At the same time, we are working toward a theory of unification of the four forces. To date, gravity has refused to be incorporated.

Various attempts by Bekenstein and others to develop (a relativistic extension) modified gravity would give much-needed credit to MOND. However, the problem of gravitational lensing and light deflection appears to be beyond MOND at the moment. Perhaps this issue may be solved in the framework of tensor-vector-scaler theory, etc. Despite its successes and achievements, a growing number of scientists have tended to dismiss it. For instance, the respected and renowned British Royal astronomer Sir Martin Rees[62] commented about MOND, on page 119 of his book entitled: *BEFORE the BEGINNING: Our Universe and Others*, that:

> *"Milgrom's proposal seems unappealing for a second reason (in the jargon of consumer magazines, it is my 'worst buy'). It jettisons one of the triumphant successes of physics—Einstein's theory of gravity, which incorporates and extends that of Newton, and has survived amazingly precise tests. The MOND idea as Milgrom realizes, would destroy the entire integrity of Einstein's theory—it is not mere tinkering (or patching up) and would set us back to a pre-Newtonian stage. That would be a high price to pay."*
>
> —REES 1997, 119

---

62 Rees, M. 1997. *BEFORE the BEGINNING: Our Universe and Others*. ISBN: 0-684-81766-7, Simon & Schuster Ltd.: Great Britain.

There is no doubt that the majority of astronomers prefer to accept a universe with DM without questioning it, rather than alter Newton's or Einstein's gravitational theory. It appears that MOND has been rejected, not for sound, objective, and scientific reasons, but because many cosmologists realize that it would put an end to the concept of DM and the standard cosmological model. So, the attempts to defend the DM hypothesis continue. Personally, I believe that Milgrom has performed a useful and excellent service and should be applauded for it by exposing the failures of the standard cosmological model—MOND predictions and successes point to the need for new physics.

Erik Verlinde's Emergent Gravity

Erik Peter Verlinde was born on the 21st January 1962, in Woudenberg, The Netherlands. His successful road to physics (with his twin brother Herman) was inspired by television shows in the 1970s about particle physics and BHs. The programs were mainly about early work on information storage in BHs by Hawking and physics Nobel laureate 't Hooft. "When I was about fifteen I saw them on television talking about the physics of elementary particles and black holes," says Verlinde. "I knew then that I wanted to work in that area." Verlinde and Herman went on to obtain their Ph.D.s from the University of Utrecht in 1988.

After graduating, they both went to Princeton, Erik to the Institute for Advanced Study and Herman to a teaching post at Princeton University. They both married and divorced sisters and Erik Verlinde returned to Amsterdam to be near his children. Currently, they both work in the exciting area of ST and their contributions to the mathematics of ST are well documented. It is worth pointing that Verlinde's early contribution is in the form of the Verlinde formula, which is widely used in conformal and topological field theory.

His second, and perhaps most significant, contribution to theoretical physics was outlined on 8 December 2009 at a symposium at the Dutch Spinoza Instituut, where he introduced a theory that derives Newton's classical mechanics. This was shortly followed by the publication of "On the Origin of Gravity and the Laws of Newton" on the 6th of January 2010. Verlinde was inspired by an incident in the south of France while on holiday. When thieves stole his laptop and keys, they could not have imagined that they had done the whole physics community a huge favor. Because of the incident, he was forced to take an extra week of vacation. It was during this additional week that he realized that gravity might not be a fundamental force of nature, but perhaps arises instead from thermodynamics.

Verlinde suggests that the key to unlocking gravity is information[63]—let us point out that Verlinde's work is similar and perhaps tied up with earlier work by Jacobsongeneral relativity", who showed that Einstein's equations of GR could be derived using thermodynamics and the HP in 1995. Obviously, Verlinde's work brings a host of various questions. Is it simply speculative or does it underline the beginning of the end of gravity as we know it? Are we ready to abandon the notion of gravity as a fundamental force? Can gravity be explained entirely by increases in entropy? What are the possible implications regarding dark energy and DM as the cornerstones of the standard cosmological model?

Many physicists are excited by Verlinde's ideas as they think he might be on the right track. There has been a feeling over the last 40 years or so, particularly since the discovery of BH thermodynamics, that we were beginning to realize that there is a special relationship between QT and gravity in the form of thermodynamics. Here we refer to the discovery of the laws of BH thermodynamics and the Bekenstein discovery of the BH entropy. The work of William Unruh, and the subsequent discoveries of the Unruh temperature and Hawking radiation, have also given us reason to believe in the intimate relationship between gravity, QT, and thermodynamics.

Verlinde's approach provides a new perspective with which to tackle head on some of the deepest issues and questions in science, particularly in physics; it may ultimately lead us to understand why time, space, and gravity exist. Einstein came close to understanding gravity when he argued that it is the

---

[63] Verlinde, E. P. 2010. "On the Origin of Gravity and the Laws of Newton." Accessed July 07, 2011. http:// arXiv: 1001.0785[SPIRES].

curvature of space-time, rather than just the curvature of space. He explained that there is no such thing as a force of gravity which pulls things to Earth, but rather that the curved paths that falling objects appear to take are an illusion brought on by our inability to perceive the underlying curvature of the space in which we live—the objects are simply moving in straight lines.

There have been several attempts in the literature to answer the questions brought up by Verlinde's paper. For instance, in 1995 Jacobsongeneral relativity"[64] showed that the Einstein equations could be derived from the laws of thermodynamics. He put forward solid arguments to explain the connection between gravitation and entropy by deriving Einstein's equations from a thermodynamics equation of state. He used the proportionality relationship of entropy and area for all local acceleration horizons. This idea has also been expanded and studied by the Indian physicist Thanu Padmanabhan[65]. For instance, attempts to provide a physical interpretation of the field equations of gravity based on a thermodynamics perspective are well illustrated in his article entitled "Thermodynamical Aspects of Gravity." Padmanabhan's arguments also lead to Newton law, where he uses Unruh's relationship between temperature and acceleration to get the gravitational acceleration. He also uses a powerful equipartition argument to provide a thermodynamics interpretation of gravity.

The works of Jacobsongeneral relativity" and Padmanabhan emphasize the role that QI (or simply, information) plays in gravity. Recently, Lee Smolin called Jacobsongeneral relativity" 's paper "one of the most important papers of the last 20 years". Padmanabhan, of the Inter-University Center for Astronomy and Astrophysics in Pune, said that the connection to thermodynamics went deeper that just Einstein's equations. He argues that "gravity is the thermodynamic limit of the statistical mechanics of 'atoms of space-time.'"

---

64 Jacobsongeneral relativity", T. 1995. "Thermodynamics of Space-Time: The Einstein Equation of State." *Physical Review Letters* 75:1260.

65 Padmanabhan, T. 2010. "Thermodynamical Aspects of Gravity: New Insights." Accessed April 12, 2015. http://arXiv:0911.5004 [gr-qc]

There have been many follow-up papers since Verlinde's publication, and an exciting new era of theoretical physics has begun. For instance, in November 2010 an excellent paper by Ram Brustein and Merav Hadad[66] completed the implementation of Jacobsongeneral relativity"'s proposal to express Einstein's equations as a thermodynamics equation of state. The paper showed that the equation of motion of generalized theories of gravity is equivalent to the thermodynamics relationship. The result of this work appears to show that gravitation on a macroscopic scale is simply a phenomenon of information. Brustein and Hadad extended Jacobsongeneral relativity"'s proof by using a more general definition of the Noether charge entropy. Before them, in 2008, E. Elizalde and P. J. Silva[67] took Jacobsongeneral relativity"'s proof to new heights by extending the simplest Einstein-Hilbert theory of gravity to more general theories in an article entitled "F(R) Gravity Equation of State".

Verlinde's paper is perhaps one of the most important of this century as it provides the missing link for understanding the origin of gravity. We have been taught that gravity is one of four fundamental forces, and most physics books describe how gravity works, but they do not explain how it arises, i.e. we do not understand its origin. According to Verlinde, the key to understanding gravity is 'information'. He showed this by deriving Newton's law of gravity from thermodynamics combined with the relationship between area and entropy. Padmanabhan reached the same conclusion years ago, as is well illustrated in many of his publications.

In the same year, Smolin showed successfully that a version of Verlinde's argument in loop QG is well supported. As rightly pointed out by Verlinde, gravity has been undressed over the last thirty years or so: it is the differences in entropy that can be the driving mechanism behind gravity, that is, as he puts it, an "entropic force." Verlinde is now actively working with his brother Herman to recast these ideas in more technical terms of ST. Many physicists

---

66 Brustein, R. and Hadad, M. 2010. "The Einstein Equations for Generalized Theories of Gravity and the Thermodynamic Relation $\delta Q = T\delta S$ are Equivalent." Accessed January 12, 2017.https://arxiv.org/abs/0903.0823

67 Elizalde, E. and Silva, P. J. 2008. "F(R) Gravity Equation of State." Accessed September 15, 2010. https://arXiv: 0804.3721v2 [hep-th]

had said it before, as Verlinde argued during a talk he gave in New York: "We have known for a long-time gravity does not exist, It is time to yell it."

Although similar ideas are found in theoretical physics, particularly in cellular automata theory, nobody has expressed it as strongly as Verlinde. He combines the laws of entropy and information theory and the physics of BHs and thermodynamics, to elegantly replicate Newton's law of gravitational attraction with beautiful mathematics, and he derives Einstein's general relativity. The work of Verlinde does suggest that gravity emerges naturally from physical dynamics, analogous to thermodynamics processes. What I have learned from Verlinde's work is that increases in entropy might entirely explain gravity, and the notion of a fundamental gravitational interaction might have to be abandoned. Verlinde's work and related arguments have raised a lot of questions and doubts about the possible non-existence of dark energy and DM.

What emerges is a realistic alternative interpretation which considers the entropy and temperature to be intrinsic to the horizon of the universe due to the information which is holographically stored there. Along the same lines, Damien Easson, Paul Frampton, and Nobel laureate George Smoot[68] suggest that we can do away with dark energy and that the accelerated expansion of the universe is due to an entropic force arising from the information storage on the horizon surface screen. The central role played by these ideas of information, holography, entropy, and temperature are emphasized. It appears that entropic forces may explain dark energy. Looking at gravity from this angle could shed light on some of the complex issues in space, time, and cosmology. For instance, dark energy and DM, believed to be anti-gravity, a force that seems to be speeding up the expansion of the universe, might have to be discarded.

An entirely different approach put forward by three Koreans physicists supports the conclusions of Verlinde's work. Lee, Kim, and Lee[69] presented a

---

[68] Easson, D. A., Frampton, P. H. and Smoot, G. F. 2010. "Entropic Accelerating Universe." Accessed June 05, 2011. http://arXiv:1002.4278v2

[69] Lee, J-W., Chan Kim, H. and Lee, J. 2010. "Gravity from Quantum Information." Accessed June 05, 2010. http://arXiv:1001.5445v2 [hep-th]

novel approach which connects gravity to information. Their paper showed how the Einstein equation could be derived using Landauer's principle, the result being that gravity has a quantum informational origin. Lee and colleagues have proposed that the erasure of QI when particles go through a holographic screen increases cosmic entropy, suggesting that gravity may originate in QI.

There is an overwhelming realization that physics is at a crossroads; we are living in exciting times considering recent advances in theoretical physics. Some relationships connect energy to matter, matter to gravity, and finally matter and energy to information or entropy. The Einstein equation links matter to gravity and matter to energy, while Landauer's principle links information to energy. We now have a relation linking information to gravity as well as one between the Einstein equation and QI. As such, it appears that Einstein's equation is more about information than energy, and there is no doubt that a new area is opening up where information is more fundamental than energy.

Jacob Bekenstein's Holographic Universe

Bekenstein's view is that our universe is like a BH with a holographic mode of operation, meaning that it behaves like a hologram. Simply put, our universe is a hologram, and we humans and our brains are holograms acting or behaving as receivers of messages or information from the holograph. In his own words, Bekenstein[70] pointed out in 2003 that: "By studying the mysterious properties of black holes, physicists have deduced absolute limits on how much information a region of space or a quantum of matter any can hold." Bekenstein's HU arose from the formulation of the entropy and energy relations of BHs. He argued that the second law of thermodynamics is violated in systems containing BHs and proposed solutions to mediate it.

---

[70] Bekenstein, J. D. 2007. "Information in the Holographic Universe." Accessed February 02, 2016. https:// www.scientificamerican.com/article/information-in-the-holographic-univ/

It took several decades for physicists and cosmologists to abandon the view that a BH is just a massive passive object that swallows everything, and that nothing near it can escape—it has nothing to do with entropy or thermodynamics. The new world view started to emerge from the early 1970s, due partially to the subsequent work of Hawking who showed that a BH does indeed emit radiation. This work showed the connections between gravitation, thermodynamics, and QT.

Further work in this area was led by a young physicist called Bekenstein. While still a graduate student of Wheeler's at Princeton, he found the conjecture that "black holes have no hair" very distressing. The principle allows one to commit a crime against the second law of thermodynamics. Drop a package containing some entropy into a stationary BH; entropy is decreased in that part of the universe far from the exterior. For an exterior observer without the inside knowledge or information about the package, it is impossible to tell exactly whether if the total entropy in the universe has decreased or not.

While thinking about these issues, Bekenstein was inspired by Demetrios Christodoulou's work investigating Penrose's efficiency of certain processes for extracting rotational energy from a Kerr BH and converting it to mechanical energy of particles. Christodoulou concluded that the most efficient processes were those associated with reversible changes of a BH. He practically showed that irreversibility limits the efficiency of processes for extracting energy from a BH, and that the most efficient processes are the reversible ones, hinting at the relevance of thermodynamics.

In his 1972 Ph.D. dissertation, Bekenstein discussed some of these issues, including the Hawking area theorem. He went on to combine the work of Christodoulou and Hawking, building on his knowledge of thermodynamics and working out a wonderful recipe for how to rescue the second law of thermodynamics from collapse. His work on BH entropy solved the paradox posed by Wheeler's demon and made possible the enunciation of the generalized second law of thermodynamics: "The sum of the black hole entropy and the ordinary thermal entropy outside black holes cannot decrease."

Bekenstein has argued that the four dimensions of our world may just be an illusion—all things in the universe are related and are part of the same unbroken continuum. The theory of BHs and entropy suggests that the information content of any region of space is defined by its surface area, not by its volume as was previously thought. As we have seen, the relationship between BH entropy and HP came in the early seventies. After several works on BHs were published, a number of researchers published a set of rules about the state and evolution of classical BHs. Inspired by these rules, Bekenstein suggested an analogy between classical BHs and thermodynamics (aforementioned).

Some physicists have come to the view that if the BH gives rise to a new universe where the information flows, unitarity is preserved in the whole of the universe but apparently lost for any observer outside of the BH. In contrast, the solution proposed by Bekenstein is elegant. In 1981 he put forward the idea of the universal entropy bound. It states that the entropy S, of a complete physical system which is asymptotically flat in space-time, has total mass-energy E, and fits inside a sphere of radius R, is necessarily bounded from above. The bound is necessary if one wants to avoid any violation of the generalized second law of thermodynamics.

Maldacena took these ideas seriously by showing that the physics inside a universe with five dimensions and shaped like a Pringle [a Pringle potato crisp] is the same as the physics taking place on the four-dimensional boundary. What Maldacena illustrated is that two universes of different dimensions and obeying different physical laws are shown to be equivalent to the HP. Following the work of Bekenstein on BH entropy, the HP was expanded and led to some theorists suggesting that gravity is not a fundamental force, but rather a consequence of the way information about objects is organized in space and time.

Following in the footsteps of Bekenstein, two distinguished physicists—Gerard 't Hooft and Leonard Susskind—have taken the HP to a new level. The presentation of the HP has changed forever the way we view how entropy or information content is stored and the counting of degrees of freedom of physical systems, and led to the realization that information plays a fundamental role in the universe.

Gerard 't Hooft and Leonard Susskind's Holographic Principle

Following the work of Bekenstein and Hawking on BH thermodynamics, which has stressed the role that information plays in physics, physicists have come to realize that information is fundamental to any theory of physics. In the 1990s the important work of Kip Thorn on the theory of BHs suggested that the three-dimensional world we know is merely a hologram projected from the furthest reaches of the cosmos. Building on Bekenstein, Hawking, and Thorn's work, 't Hooft[71] developed what is known today as the HP. Susskind extended and added a new dimension to this concept within ST.

In 1993 't Hooft[72] published a paper entitled "Dimensional Reduction in Quantum Gravity," in which he outlined the basis for the concept and pointed out the closeness of the theories of QG to holograms). A year later, in September 1994, Susskind published "The World as a Hologram," which added a new dimension to 't Hooft's work and extended the concept of the hologram in STs. He describes the two-dimensional plane of 't Hooft as a screen onto which all the physics in the three-dimensional world is projected. He extended this idea to ST and made a direct comparison by pointing out that ST in a light cone gauge seems to be a realisation of the same concept, but where one of the dimensions mysteriously disappears.

Gerard 't Hooft's HP states that: all the fundamental information in one part of the universe can be equivalent to some information (or physics) defined on the boundary of that part of the universe. For instance, we know that the maximum entropy or information content of any region in space is defined by its surface area rather than by its volume. With knowledge of BH

---

[71] 't Hooft, G. 2000. "The Holographic Principle." Accessed September 02, 2010. http://arXiv: hep-th/0003004v2]

[72] 't Hooft, G. 1993. "Dimensional Reduction in Quantum Gravity." Accessed October 11, 2012. http://arXiv:gr-qc/9310026.

thermodynamics, we can calculate the absolute limits of the holographic bound. This holographic bound is simply the amount of information a specified region of space contains (or simply how much information can be included in a region of space). The concept is best exemplified by the Bekenstein bound on the maximum entropy in each region of space. Susskind pointed out that, in a certain sense, the world is two-dimensional and not three-dimensional as previously supposed. Although the world around us has three spatial dimensions, the HP says that its contents can fit, or simply be encoded, on a two-dimensional surface. The HP proposal put forward by 't Hooft is well illustrated by two basic assertions:

- All the physics which takes place in some region of space can be represented by a theory which lives on the boundary of that region (a hologram encoded in a section of 3D space as a light interference pattern on a 2D surface). To put it in another way, information from a (D+1)-dimensional space is encoded onto a D-dimensional space.
- The theory on the boundary of the region of space should contain at most one degree of freedom per Planck area.

HOLOGRAPHY AND THE HOLONOMIC MODEL OF THE BRAIN

Karl Pribham's Holographic Universe

Before the advent of QT and the holonomic model of the brain, our worldview was dominated and heavily influenced by Newtonian (classical) models of physics. This mechanistic view of the universe has led to tremendous progress in all areas of human life. It has been partially responsible for the industrial revolution, the development of science and technology, and has had great influence in in psychology, history, etc. Moreover, this has indeed dominated western culture and our ways of thinking and living for centuries. To some extent it has influenced the rest of the world. As a result, almost all areas of knowledge have been modeled on the dominant paradigm.

An example is the mechanistic view that saw the brain and mind as a machine. Since the discovery of QM this worldview is slowly evaporating.

A new paradigm led by Einstein, and extended by Bohm, Pribham, and a growing number of digital physicists, is revealing to us an entirely different universe with worldview distinct from the mechanical Newtonian one. Despite the advent of QT, and coupled to the rise and domination of computers, the brain came to be regarded as a computing machine[73].

Roger Penrose[74] in particular has shown that the mind could not be an algorithmic process. Several others, such as Kampis and Pattee, have shown in their work that any support for the computational model is untenable. Likewise, Pribham has shown that the paradigm based on the computer model of the brain is flawed. Despite the failure of the computing model to account fully for how the human brain works, it has had some success at explaining aspects of the brain's working from the approach of neural networks. However, there are other processes in the brain that cannot be accounted for by such models as they clearly fall outside the neural net. The overwhelming consensus seems to be that brain processing cannot be described satisfactorily by Turing machines[75].

The key to the development of a new approach would come from the elaboration of holography, which can be traced from earlier work by the mathematician Leibniz. In the eighteenth century, the French mathematician

---

[73] As evidence against it mounted, it did not take long before the conventionally-held view that considered the brain as a computational device started to crumble. This is because the approach adopted for the description of the brain as a computer (and many related scientific publications) had not kept pace with developments in the physical sciences, particularly QT. Recently, QT has gained ground in other areas of learning so that some brain scientists have begun to take seriously the quantum-like operations in the brain. This new approach has been seriously considered as a result of the setbacks suffered by previous models, mainly various computational and artificial intelligence models. As such, a growing number of scholars have shown the shortcomings of the computational model of the brain and have in fact pointed out several of its limitations.

[74] Penrose, R. 2007. *The Large, the Small and the Human Mind.* Cambridge University Press: Cambridge.

[75] The number of fundamental questions which had been left unanswered by the various computer models was a sure sign that it was time to look at new approaches. The binding of patterns, the inability of machines to match computing capabilities on humans and non-humans, and lastly, the inability of computers to show the

Fourier explained the mathematics behind the harmonic wave resulting from the interference pattern of waves. Denis Gabor, building on the work of his predecessors, made immense contributions to the mathematical principles of holography; he then applied the work of Fourier, in particular Fourier transformations, to the creation of the hologram. Gabor's work was never intended to be used in brain research, but was instead meant to find "the compressibility of a telephone message that could be sent over the Atlantic cable without destroying its intelligibility."

The Gabor function thus describes both units of brain processing and of communication. The realization that the same mathematical principle that guides the brain also describes communication theory was a landmark. The Gabor quanta of information have shed so much light on the holonomic model. By 1947 Gabor's immense contributions to holography came to be acknowledged by many, and with the invention of the laser in 1960 (by Bassov, Prokhorov, and Townes), brain researchers had a powerful tool in their hands. One of the most prominent researchers at the time took advantage of this concept: Pribham linked the hologram and holographic brain functions and proposed a holographic model of the human brain, suggesting that the human brain can be regarded as a hologram of the entire universe.

Between 1972 and 1982 Bohm promoted the HU, suggesting that every part contains the image of the whole. Later on the combined work of Bohm and Pribham came to be known as the holographic paradigm, which argued that everything around us is merely a holographic projection of one universal consciousness. In 1982 Aspect's[76] experiment, where he showed that two photons from one atom always communicate instantly regardless of their separation, gave a tremendous boost to Bohm's theory of unbroken wholeness.

---

self-organization found in biological systems; all suggested that a whole new line of reasoning was needed.

76 Aspect, A., Dalibard, J. and Roger, G. 1982. "Experimental Test of Bell's Inequalities using Time Varying Analyzers." *Physical Review Letters* 49:1804. Accessed 1st May 2013 http://www.phys.ens.fr/~dalibard/ publications/Bell_test_1980.pdf

It appears that by combining QM, holography, information, and the brain we are gifted with a new and rich paradigm named the *quantum holographic informational universe*. A quantum holographic informational system is interconnected by a universal non-local information flux. Non-locality is a fundamental property that has been shown to exist at the quantum level and at the macroscopic one; it is behind all instantaneous interactions and interconnections between all parts of the universe. Non-locality has been proven to exist in quantum physics. For instance, the work of Aspect and colleagues demonstrated the presence of non-local action at a distance between two photons emitted by one atom[77].

The Holonomic Brain Theory

Albert Whitehead, one of the strong opponents of dualism, emerged in the late 1920s to refute the claims against the dualism of mind and matter. A mathematician and philosopher, he argued that reality is inclusive and inter-related. In the same spirit, a neurosurgeon called Karl Lashley[78] published a number of papers which showed that memory is not located in any particular part of the brain, but is distributed throughout it. The holographic nature of reality is backed by the earlier work of brain scientist Karl Lansley in the 1920s. A series of experiments on rat brains showed that, regardless of the amount of a rat's brain that is removed, there is no effect on the ability to perform complex tasks learned before the surgery. Experiments conducted by many researchers show that optical and other memories do not have to be stored—and are not stored—in any specific location in the brain.

---

[77] Work by Gisin and colleagues in 1997 showed the existence of instantaneous transmission (non-local quantum) actions on a macroscopic scale. The synthesis of quantum models with information theory and self-organization provide us with the ability to explain many of puzzles of brain function and behavior. How does brain function mirror a quantum process in physics? The combined works of Bohm, Pribham, and Aspect demonstrate that the brain functions as a quantum system and prompts the realization in quantum theory that each particle of the universe contains all the information present in the entire cosmos.

[78] Lashley, K. 1950. "In Search of the Engram." In *Physiological Mechanisms in Animal Behavior*. Academic Press: New York. pp. 454–482.

Insight by Gabor led to the use of windowed Fourier processes for use in communication theory. He went on to call his units of communication quanta of information in an approach similar to that used to describe quantum processes in particle physics. He used Leibniz's calculus to develop the concept of holography. He received the Nobel Prize for his work in constructing the first hologram. It took years before the discovery of the mathematical principle of holography (based on Leibniz's calculations and Fourier mathematics) could be proven and demonstrated experimentally by the discovery of the laser.

The paradigm shift from the mechanical viewpoint to a more holistic approach is mainly due to three significant developments: the discovery of the hologram, the emergence of QT, and accumulated experimental evidence in brain research in the 1960s and 1970s, particularly the work of Stephen Kuffler which mapped certain brain processes as local field potentials. In 1965, Emmet Leith and Juris Upatnicks constructed holograms based on the newly invented laser beam. Extending the work of Lashley, Pribham proposed that the brain is a hologram. His experiments showed that the structure of the human brain is holographic. The activities of the brain transcend time and space.

In 1971 Bohm put forward the hypothesis that the universe may be like a giant hologram, thus hinting at the possibility that it may help to explain many of the unsolved puzzles of physics. Michael Talbot sums up some of the ideas of Bohm and Pribham in his book *The Holographic Universe*, by arguing that:

> *"Our brains mathematically construct objective reality by interpreting frequencies that are ultimately projections from another dimension, a deeper order of existence that is beyond both space and time: The brain is a hologram enfolded in a holographic universe. The brain converts waves and frequencies to make reality look concrete to us."*
>
> —TALBOT 1992, 54

What is the meaning of the word holonomic? It describes a constrained, windowed, Fourier process. Holonomic processes are known in physics as quantum holography. Holonomic brain theory refers to a process that occurs in fine-fibered neural webs. According to Pribham's holonomic theory, the receptive dendritic fields in sensory cortexes are described mathematically by Gabor functions. This can be illustrated by how our brains process images. The retina first transforms each image we see via a quantum process, which is in turn transmitted to the visual cortex. The cortical receptive fields form patches of dendritic local field potentials which are described mathematically by Gabor functions.

Gabor functions fail to serve as the properties of images that guide us and allow us to navigate around the world. Many of these theories of perception based on frequencies gave rise to a transformational view of the processing of visual signals. Moreover, this transformational view rests on the Fourier theorem. Therefore, the Fourier approach to sensory perception constitutes the basis for the holonomic theory of brain function. The Fourier transform (FT) formula converts spatial forms to the frequency domain. Holonomic theory is well illustrated and best understood through the mathematical notion of FTs. A FT is a special case of a wavelet transform with basis vectors defined by trigonometric functions (sine and cosine); for example, a particular sound wave contains a particular frequency. The FT algorithm uses discrete FTs to represent a fractal, for instance, a stone in the frequency domain.

Using FTs, a function can be expressed as a sum of sine or cosine waves at different frequencies. What the FT does is simply convert the function to the frequency domain. A FT converts a function of time f(t) into a function of frequency F(jw), where w is angular frequency. Alternatively, the FT can convert spatial coordinates to the spatial frequency domain. As shown by Pribham, if the brain performs a FT for visual stimuli it may be possible that it can also perform a FT for other senses such as hearing, taste, or smell. It is interesting to note that the mathematics of the FT are independent of the data sets.

In 1979 some experiments were conceived and performed by De Valois[79] and colleagues on animals (cats and monkeys) mainly to find out if the cortical cells responded to differences in the Fourier spectra. Astoundingly, the results showed that the visual cortical cells respond to the Fourier fundamentals. Furthermore, experiments conducted by Pribham in 1994 with the rat somatic sensory system showed that cortical cells were also found to respond to spectral information. According to Pribham, temporal and spectral information are simultaneously stored in the brain. From the holonomic brain theory viewpoint, our brain is continuously engaged in correlation processes and this is how the senses are integrated.

Advances in holonomic theory claim that the act of thinking and remembering is concurrent to the action of the inverse transform. For instance, regarding vision it argues that the image of an object formed on the retina is converted into a holographic or spectrum domain. In turn, the information in this spectral holographic domain is distributed over an area of the brain by the polarization of various synaptic junctions in the dendritic structures. Understanding this process means that there is no such thing as a localized image stored in the brain. Conscious awareness and memory are simply the product of the transformation from the spectral holonomic domain back to the image domain and vice-versa.

Pribham's experiments showed that even if certain parts of the brain are damaged, no memories are erased because memories are simply restored as frequencies. Further support for the holonomic brain can be understood through the dissipative structure model proposed by Pribham. We learn from him that dissipative structures self-organize around a different principle called 'least action principle'. Accordingly, in the brain the entropy is being minimized (which maximizes the amount of information possible to store) according to the least action principle. Thus, the system (brain) self-organizes such that more and more information may be retained.

What are then the parallels and similarities between how the brain processes information and quantum processes? Billions of single cells operate in the

---

79 DeValois, K., DeValois, R. and Yund, W. 1979. "Responses of Striate Cortex Cells to Grating and Checkerboard Patterns." *Journal of Physiology* 291:483–505.

brain. It has been shown mathematically that the functional interactions between single-cell processes and the branches from the single cells are very similar to the descriptions of quantum events. There is no doubt that there are some similarities between how the brain works and basic principles operating at the quantum level. Furthermore, the mathematics used to model how the brain network operates is due to earlier work by Hilbert, which Heisenberg then borrowed and extended to QM. Some credit should also go to Denis Gabor, who introduced these ideas first to psychophysics and later to holonomic theory.

The dendrons in Figure 2.4 are the sites where brain processing occurs; they are sensory receptive fields. From Pribham's holonomic theory we learn that the cerebral cortex is the location of holographic information processes, known as multiplex neural holograms, which are dependent on local circuits of neurons. Accordingly, these neurons function in the undulatory mode and are responsible for the horizontal layer connections of the neural tissue where holographic interference patterns can be built. Pribham argued for their fundamental importance in 1991 [quoted by Di Biase[80]], by stating that:

> *"...as sensory generated receptive fields, they can be mapped in terms of wavelets, or wavelet-like patterns such as Gabor Elementary Functions. Denis Gabor called these units quanta of information. The reason for this name is that Gabor used the same mathematics to describe his units as had Heisenberg in describing the units of quantum microphysics. Here they define the unit structure of processes occurring in the material brain..."*
>
> —DI BIASE 2009, 661

This statement by Pribham supports the idea of quantum holographic interactions and the universal interconnectedness within the brain, the universe, and the cosmos; it is instantaneous (a non-local connection) like Bohm's

[80] Di Biase, F. 2009. "Quantum-Holographic Informational Consciousness." *NeuroQuantology* 7(4):657–664.

theory of wholeness. The work of Pribham demonstrates beyond any doubt that memories are not encoded in a specific location like neurons, but in patterns of nerve impulses, suggesting that the cerebrum behaves like a hologram. If a brain behaves like a hologram, this explains why the human brain can store huge amounts of information and memories in little space. It is well documented that the human brain can memorize on the order of 10 billion bits of information during a lifetime. This is like a hologram which has the capacity to store a significant amount of information.

Advances in QT and other studies support the unifying paradigm of the non-local carrier of information. In 1995 Ervin Laszlo[81] argued that memory maybe stored in a collective holographic memory field that lies outside the boundaries of the human physical body. Our brain then can access the memories from that field and those memories are not stored within the brain. Supporting Laszlo's view, recent advances in brain research have lent credibility to vast amounts of memory storage capability outside the physical body.

Experiments have shown that one cubic centimeter of film in a hologram can hold up to 10 billion bits of information. It is possible to argue that our ability to retrieve information from our memories may only be explained if the brain functions according to HP. Consciousness is dependent on the holographic structure of the brain and memory, and it appears that to understand consciousness and the workings of the universe we may have to rely on a holographic model. Memory and consciousness are not stored specifically in one place in this universe, but memory patterns are like waves that permeate the whole universe.

Similarities between the brain and the hologram illustrated in Table 2-2 may help us to make sense of viewing the brain and the universe as holographic systems.

[81] Laszlo, E. 1995. *The Interconnected Universe: Conceptual Foundations of Trans-Disciplinary Unified Theory*. World Scientific Publishing: Singapore.

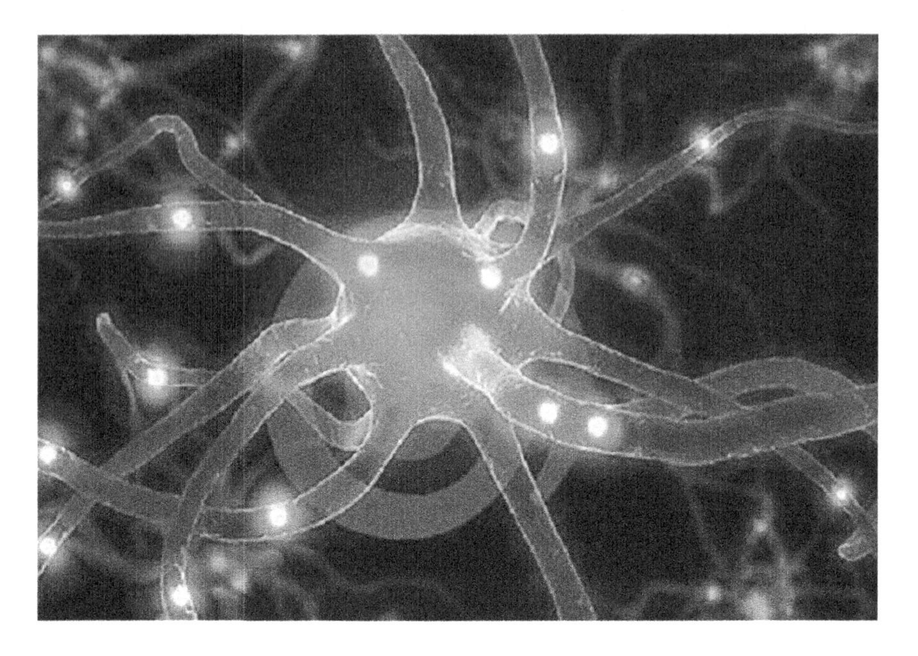

Figure 2.4 *Dendrons (dendrites) are sensory receptive fields where brain processing occurs.*

Table 2.2 *Similarities between the brain and holograms.*

The Brain	Holograms
Memories are encoded in patterns: nerve impulses that crisscross the entire brain but are not encoded in neurons.	Patterns of laser-light interference crisscross a piece of film that contains a holographic image.
The human brain is capable of memorizing up to ten billion bits of information.	Use information storage that is capable of recording many images on the same surface: one cubic centimeter can hold up to ten billion bits of information.
The information in our brain appears to be instantly inter-related to every other piece of information.	Every part of a hologram is interconnected with every other part.
The brain can translate frequencies that it receives via the senses (light frequencies, sound, etc.) into concrete perceptions.	A hologram uses a process of encoding and decoding frequencies: the hologram functions as a lens that can convert various frequencies into a coherent image.

The Brain	Holograms
The brain uses holographic principles to perform its various operations.	Holograms use holographic principles to convert frequencies that are received through the senses mathematically. The nature of the universe could be like a hologram.
Information is not localized in the brain but rather is spread throughout the whole brain. Each brain part in some way contains the whole memory. The brain has a parallel-processing capability; it can recognize words and objects by association and can retrieve information.	Information is not localized in holograms; instead, information is spread throughout the hologram as a whole and not in parts. This is why one can never have half an image only, it is always the whole picture. If a hologram is sliced up, each fragment will continue to contain the whole image: that is, the entire object in smaller and smaller fragments.
Humans can locate the source of sounds without moving their heads, even if they can only hear with one ear. This process can only be explained through holographic principles.	Holograms, like the brain, can recognize words and objects by association. For instance, a hologram can be used to scan a page; when it detects a target word or image, it lights up in a way that is highly similar to how our brains enable us to recognize words, faces, and everything else around us.

To sum up, I would like to point out that, from the holonomic theory of Pribham, we have learned that the brain functions, processes information, and behaves like a hologram and has access to the larger universal hologram. The work of Pribham, Bohm, and Gabor is well supported by the research of John Bell[82] in QT who, in 1964, showed that subatomic particles are interconnected in a way that transcends time and space. Whatever happens to one particle affects all other particles. Bell's experiment is well documented and goes beyond Einstein and the wave-particle theory.

The theory of the non-local QH has been developed by several researchers but emerges particularly from the work of Bohm, Pribham, Edgar Mitchell, Marcer, and Walter Schempp[83]. The process of the QH is based on the

---

82 Bell, J. S. 1964. "On the Einstein Podolslosky Rosen Paradox." *Physics* 1:195.

83 Marcer, P. J. and Schempp, W. 1997. "Model of the Neuron Working by Quantum Holography." *Informatica* 21:519–34.

principles of quantum holography and quantum entanglement (interconnectedness at the quantum level) to achieve non-local information. The theory combines the science of quantum entanglement (interconnectedness) of objects and events at the quantum level and holography, particularly the two most important aspects of it—the information processing property and the storage capacity.

Evidence accumulated over the years suggests that human beings perceive in two ways: by local and non-local processes. Our eyes respond to electromagnetic waves. Recent findings from many sources indicate that our mind also responds to some non-local communication. Three essential steps take place during quantum coherence processes between particles (non-local communication). Firstly, the interaction is instantaneous; secondly, the interaction is independent of the distance between the particles; and thirdly, the interaction is impervious to shielding. In a nutshell, non-locality refers to a process in which signals propagate instantaneously regardless of distance.

QH teaches us that information transmission can occur through any energy wave field, i.e., quantum, electromagnetic, mechanic, acoustic, and others. In a nutshell, from the smallest entity of the quantum domain to the macro-scale, all objects and entities in the universe are made up of energy and in constant oscillation at different energy frequencies. The objects emit or generate energy fields that interact with other objects. The QH paradigm is about the relationship between matter, energy, and information. Every object or particle emits and has a frequency; it is continuously broadcasting.

From the holographic model of consciousness we learn that consciousness is not stored in any specific location in the brain, but is scattered throughout the brain. The cerebral matter works by analyzing sensory information through mathematical analysis of temporal and spatial frequencies. It is possible that the universe is organized in the frequency domain, where there is no space or time.

The two most important aspects of the hologram are its three-dimensionality and its wholeness. The three-dimensionality is what led Pribham to argue that the brain does not store memories in one specific location, but rather distributed throughout the brain. We have learned that it does not matter what part of the brain is damaged, memories are not eradicated because of these multiple distributions. The second important feature, wholeness, offers a possible viable alternative explanation for why memories in the brain are not located in a specific place in the brain. For instance, if a holographic plate is broken into tiny fragments, each tiny piece of the holographic film contains all the information of the whole. It is possible to reconstruct the entire image from one tiny piece. The brain is a hologram—the holographic view of the brain and the universe is a reality.

PART THREE

Fractal Holography: Space-Time and Cosmology

Fractal Holography

If the universe is built up from logic and law then time is symmetric but if the universe is built up from consciousness then the computational geometry will have a fractal quality.

—R.A.DELMONICO

FRACTALS AS A SIGNATURE OF A HOLOGRAPHIC UNIVERSE

Despite all the recent fundamental research and scientific breakthroughs in cosmology, physicists have not yet been able to give a satisfactory answer to the basic question: what is the origin of fractals? Included in this issue are various sub-questions that are also not yet answered. They include the following:

- Why do fractals occur in our universe and perhaps others?
- How did fractals come into existence?
- What is the relationship between matter, energy, and consciousness?
- Why are fractals everywhere?

As this concept is of fundamental importance and will further illuminate our understanding of the idea of holography, and because of its novelty and broad possible range of application in various scientific fields, it deserves to be thoroughly explored. Here, we examine its origin and attempt to explain it with regard to matter-energy equivalence, information, frequencies, consciousness, and the structure of the universe. There exist out there a lot of books and websites covering fractals. My intention is not to repeat what has already been written, but simply to look at it from a different point of view and make connections with the field of digital physics.

I will also discuss the fractal nature of the hologram by exploring the concept of information attached to it to reveal the interconnection between information, space, time, and matter. This can only stress the important role it plays in the field of digital physics. Fractal, from the Latin fract(us), i.e., fraction of something, is a word that was coined by the French mathematician Benoit Mandelbrot[84] in 1975. Born in Poland, Warsaw, to a family focused on academia, he immigrated to France in 1936 following the German threat to attack Poland. Fractals are all over our universe, and in fact, most of the natural objects we come across and observe daily, such as trees, brains, stones, mountains, clouds, grasses, etc., are fractals.

Any fractal we find in nature is a textured geometric shape—when split into tiny parts, each one of those parts is roughly a smaller version of the larger part. Almost all fractals are a rough, complex, or fragmented geometric shape. Mandelbrot defines a fractal as "a rough or fragmented geometric shape that can be subdivided into parts, each of which is (at least approximately) a reduced-size copy of the whole." Some fractal properties include:

- fine and complex structure,
- intricate,
- defined by a recursive process,
- too irregular to be described by traditional geometry,
- self-similarity,
- fractal dimensions.

Although Mandelbrot made immense contributions to the understanding of fractals, it is worth pointing out that Gaston Julia is rightly considered to be the father of fractals. His early work in the 1900s reveal some of his amazing contributions to iterated functions, in addition to the Julia sets which he drew by hand. Around the same time, other researchers working on similar topics

[84] Mandelbrot, B. 1977. *The Fractal Geometry of Nature*. W. H. Freeman and Company: New York.

made an impact on the development of fractals, for example, Sierpinski's triangle and Koch's curve. The record shows that research on fractals continued until the 1960s, but interest in the subject waned due to lack of progress, possibly in turn due to a lack of new mathematics and applications.

The advent and development of computers and related subjects, such as computer graphics in the 1970s, led to the re-emergence of fractal research. This trend has continued to the present day. Today, computer generated fractals offer us some of the most successful tools for simulating nature. Further developments in this field had to wait until Mandelbrot took advantage of the computer to create what became to be known as the Mandelbrot set. In 1977, he built upon the research done by the French mathematician Gaston Julia and others to create a computer-generated fractal. He created his set by putting a simple formula into his computer and allowing it to replicate over and over again. The result is an amazing infinite amount of graphic that can be explored forever—it shows self-repeating patterns and reflects organic objects in the infinite amount of complex shapes, colors, etc. It stands out as a truly amazing discovery. The Mandelbrot set is illustrated in Figure 3.1.

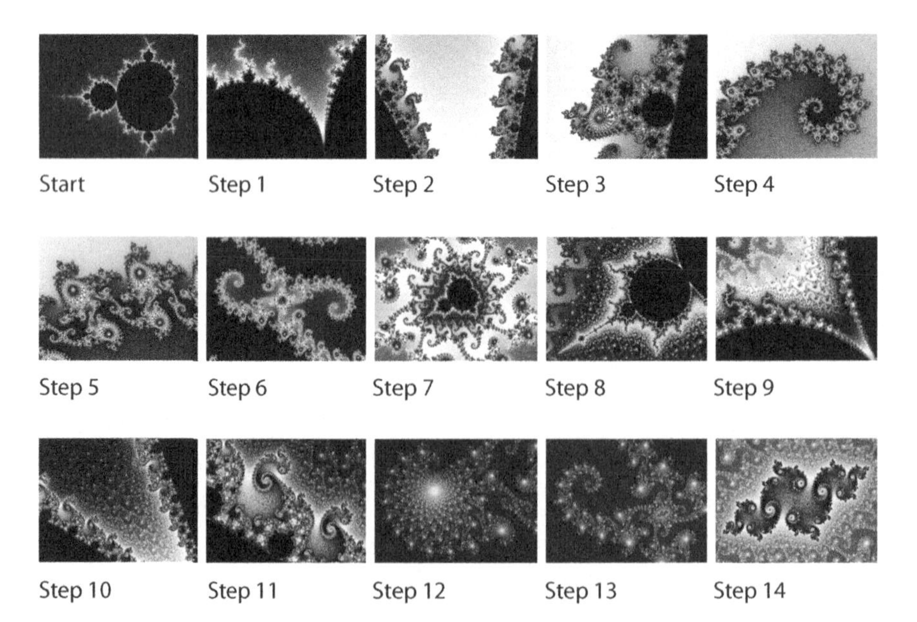

Figure 3.1 *The Mandelbrot set.*

Closer observation and analysis of a fractal such as the human body shows how different levels of nature are connected and separated. Fractal geometry can also be viewed as a bridge linking the inner (inside) and outer (outside) of everything around us. For instance, in a human body we find many areas of communication and transportation amongst various organs and systems as well as between the body and the outside world. Blood can easily circulate within our body through the fractal branching of arteries and veins.

In Africa, fractals occur in every area of life from the designer clothes of men, women, and children. For instance, people who live in the same village, tribe, or family display the same or self-similar behaviors. Researchers have found that some of these fractal designs are highly intuitive, while others are based on an advanced algorithm where pieces follow the rules and are built recursively. Fractals thus give us a new way to think about our complex universe, and to some degree help us to understand the links that bind us together. We can see how each part of the universe is related to the other parts and vice-versa.

There is a realization that everything in the universe, from the structure of our DNA to the formation of planets, galaxies, the world, the body, the brain, etc., follows the reoccurring patterns of fractal geometry. From the formation of mountains, rivers, and planets to the growth of plants, as well as in the structure of the human body, we can see a similar structure reoccurring in patterns from a tiny, smaller, simpler form to the larger, complex body. Everything in nature is simply the reflection of the whole picture. For instance, there is a similarity between the structure of an atom with orbiting electrons and the Sun with its planets orbiting around it. The world is made up of continents, countries, cities, towns, villages, etc. and in turn, villages are made up of populated families, which are made up of individuals.

This interconnectedness aspect of fractals can be viewed in the structure of many objects in nature—cauliflower and broccoli are two great testaments. A careful analysis of cauliflower shows that the flowers have the same structure throughout; this structure, regardless of size, is the same throughout the whole head. It seems that we are all interdependent and linked to one another, inseparably connected to everything around us, to our planet and the universe. This interconnectedness hints at the possibility that the human race, as well as the universe, is fractal in nature.

Let us now examine if there are any fundamental similarities between holograms and fractals. One of the most important properties of a hologram is wholeness. Any part of a hologram contains the whole image. If you cut one small part of the hologram, that part contains the whole picture—hologram and fractal are the two sides of the same coin. Each cell in our bodies contains all the DNA information to create an identical copy of the whole organism—similar to fractals and holograms. The whole is in each part and vice-versa.

The fractal nature of holograms is undeniable; both fractal and hologram have a common attribute called wholeness. There are also other similarities, thus leading us to conclude that our reality is a hologram. What we have learned from fractal geometry and holograms suggests that everything we see and observe around us, such as our solar system, planet, world, trees, people, atoms, and electrons, etc. are just smaller representations of the whole.

I would like to point out that fractals have become an active area of research and are having a greater impact in so many fields of the sciences and arts, for instance. Examples include:

- In computer graphics: many artists can create realistic landscapes using fractal equations which iterate to generate complex images, hence removing the need for high volumes of data storage.
- In biology and medicine: researchers can measure the irregular shape of red blood cells to make medicines that won't harm human beings.
- In physics: engineers have designed antennas shaped like fractals that are small enough to fit inside a small wireless phone and are capable of carrying strong signals.
- In mathematics: fractal equations are used in the wildly random world of financial trading, stock markets, weather patterns, etc.

Fractal Geometry

Fractals are objects that defy conventional measurement—it not possible to measure them with any conventional apparatus. They are crinkly objects of complicated shapes, but with a very fine structure and a high degree of geometrical complexity that defies conventional measurement. For example, cloud dimensions far exceed any topological dimension. Consequently, the concept of a whole number does not apply to the length, area, or volume of a fractal; their fractal dimension often characterizes them. What is a fractal dimension and how does one determine the fractal dimension of an object? Let us first look at some geometrical shapes that we can find in our daily life: a point, line, plane, square, or rectangle.

What all of these shapes have in common is their dimension, which is a whole number. In Euclidean mathematics, a point has zero dimensions or no dimension at all; a line has one dimension; a plane is two-dimensional; squares and rectangles are two-dimensional, and a solid cube is a three-dimensional object. These objects are characterized by their topological dimension, which is an integer number.

Euclidean geometry deals with regular shapes such as points, squares, lines, circles, triangles, etc. In contrast, fractal geometry describes the irregular shape and deals with mainly rough and uneven shapes. These irregular-shaped objects are self-similar, which make them difficult to measure; for instance, the self-similar structure of our brains, rocks, trees, mountains, or snowflakes, etc. Nature abounds with many complex shapes. Mathematicians largely ignored most of those objects which have bizarre shapes because of their complexity. What Mandelbrot did was to invent a new mathematics, called fractal geometry, to study these irregular complex shapes.

In fractal geometry, complex geometric objects have fractional dimensions and a fractal has a fractional dimension. It could be 1.2, 2.1, 2.2 or 2.3-dimensional, etc. Since Einstein's relativity work, the concept of four dimensions (four-dimensional space-time) has gained acceptance, so that time, for instance, is accepted as the fourth dimension. (We live in three dimensions plus one, space-time.) Those who work in superstring theory believe that there are up to 26 dimensions in our universe.

With the advent of computers, the fractal dimension can easily be generated by computer graphics. The fractal dimension is a fractional value that describes how irregular an object is and how much of the space it occupies. Various methods to establish the fractal dimension of an object have been proposed; the box-counting method, the mass method radius dimension, and the Bouligand-Minkowski method are well documented. The fractal dimension is best understood through the idea of scaling or magnification.

The idea of scaling is illustrated with the three objects shown in Figure 3.2, namely a line, a square, and a rectangle. Let us scale everything in the figure by a factor of ½ or 0.5. The line becomes ½ of itself, the square becomes ¼ of itself, and the rectangle also becomes ¼ of itself.

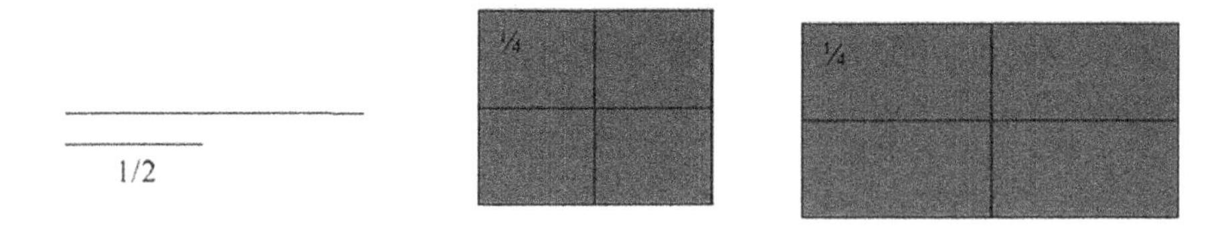

Figure 3.2 *Illustrative example of scaling a line, a square, and a rectangle by a factor of ½.*

For instance, scaling a line by a factor of **½** gives us an exact (reduced copy) of the original set. It is worth pointing out that not all fractals have a fractional dimension. Fractals are characterized by two properties: self-similarity and semi-self similarity. For instance, Sierpinski's triangle in Figure 3.3 and the Koch snowflake are self-similar, as each can be expressed as a union of sets; each is exactly a reduced copy of the full set or the whole.

Not all fractals we see in nature display this precise form. Many fractals are not exactly a union of reduced copies of the whole; however, they tend to have the same qualitative appearance. For instance, mountain coastlines display the same qualitative appearance or property throughout; they are regarded as statistically self-similar or semi-self-similar.

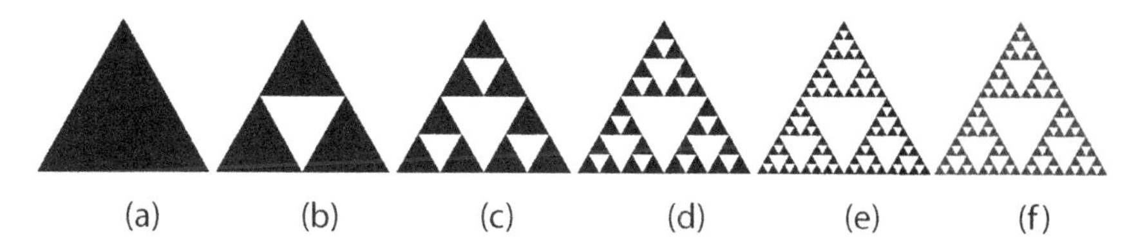

Figure 3.3 *A typical example of a self-similar fractal: the Sierpinski Triangle.*

The steps required to go from (a) to (f) are described below:

(a) Draw a triangle of any size.

(b) Draw the lines connecting the midpoints of the sides: three new triangles look exactly like the original triangle.

(c) Repeat the operation as above.

(d) The result of the above iteration is the triangle (d).

(e) The iteration of (d) has the form (e).

(f) Iterate in this way forever. The resulting figures consist of three separate identical miniature pieces.

Fractals and Chaos Theory

The emergence of chaos in complexity theory is characterized by the high presence of fractals. What is the origin and meaning of chaos? The word chaos comes from the fact that systems that the theory describes are apparently disordered, not ordered. Moreover, the idea is to find the underlying order behind the apparently random data. Most of the traditional science subjects such as physics, chemistry, or biology deal directly with predictable phenomena such as gravity, mass, power, electricity, chemical reactions, the human body, etc. Chaos (or the study of nonlinear systems), on the other hand, deals with nonlinear things that are impossible to predict or control; for instance, turbulence, the stock market, or the weather, etc. These phenomena are often described by fractal mathematics which, as we have seen, captures the infinite beauty and complexity of nature. We have come to learn from chaos theory that systems we once thought to be completely chaotic have predictable patterns.

The first experiment on chaos theory took place in the 1960s, when a meteorologist named Edward Lorenz, then at MIT, was working on the problem of weather behavior and prediction. He designed computer models of the weather with twelve equations; one hour after performing the experiment he checked the data and remarked that the result had changed dramatically and the sequence had evolved differently. Instead of the same results and patterns as before, the data had diverged: the result was different from the original one.

The experiment led him to make the major discovery that the system never predicted the weather but instead predicted the likely outcome of what the weather might be. He concluded that it is impossible to predict the weather accurately. His experiments (and other related works in meteorology), led him and others to discover many of the aspects of what is known today as chaos theory. The chaotic fractal nature of our universe can give us insight into everything around us. We learn from chaos theory, and the closely related field of fractal geometry, that many events that were considered to be chaotic, unpredictable, and random such as the weather have been found to be otherwise. Behind this complex and chaotic behavior there is a real form and pattern hidden within the chaos or randomness.

Many scholars believe that everything in our universe (planets, galaxies, stars, weather) seems to have been placed randomly. However, some mathematicians have rejected the notion of randomness, arguing that there is a chaotic equation to describe any randomness. It follows that we can describe those behaviors by using fractals and chaotic patterns. Moreover, with fractal geometry, we may use fractal types of equations to solve the random.

What is the common attribute between chaos and fractals? In relation to chaos theory, fractals are the models generated by mathematical equations resulting in chaotic systems. Fractals and chaos are used to model the structure of circulatory systems such as arteries, lung, veins, heart, brains, DNA, and nerve networks, etc. The interesting link between chaos and fractals appears to be in their properties of self-similarity, symmetry, and complexity. Mathematicians also believe that complex patterns such as fractals may find useful applications in predicting complex and random things such as the weather. For instance, a random fluctuation in the weather over five or six years cannot be represented by a simple linear equation.

> *In biology, fractals and chaos are used to accurately model and describe various parts of the human body such as the brain. The record shows that chaos theory applied to biology started in the early 1970s when some researchers were curious to implement the theory to model population trends. However, it was only after Mandelbrot published his work (the Mandelbrot set) that some researchers in biology actively started to research and identify diseases that show chaotic behaviors. This line of inquiry on DNA demonstrated that its sequences could be very like fractals because they display qualities of Brownian self-similarity.*

Fractals and Consciousness

Consciousness remains one of the least understood enigmatic properties of the human mind. The ability of consciousness to generate an inner imagination and view of the outer world has remained elusive; this thinking thing inside our mind continues to puzzle researchers and has defied any scientific or experimental investigation. The nature, as well as the location, of consciousness in human beings remains a mystery since ancient times. These difficulties have been pointed out by many researchers[85] and are well documented; one of them is the binding problem[86].

The failure of appropriate scientific approaches to analyzing the human consciousness has led many researchers to adopt a reductionist model of

---

85 Chalmers, D. 2010. *The Characters of Consciousness*. Oxford University Press: Oxford.

86 The binding problem is perhaps the most complex part of the mystery of consciousness. How is it possible that multiple memories and different impulses combine simultaneously to produce a moment of conscious awareness when data are scattered throughout the brain and there appears to be no central station or processing unit to coordinate the information?

electro-physiological mechanics of the mind. Many of these reductionist computational models of the brain are successful in explaining the computation of nerve signals in part of the brain, but have so far been unable to explain the conscious experience emerging from nerve signals. Possible lines of reasoning and answers that could lead to a solution for this puzzle have already been explored in the form of the holonomic brain theory. The fractal structure provides an underlying model in which consciousness is understood.

Erhard Bieberich[87] has developed a successful fractal approach to the integration of information perceived in the consciousness model; a detailed explanation can be found in his excellent article entitled "The Structure in Human Consciousness: A Fractal Approach to the Topology of the Self Perceiving an Outer World in an Inner Space." Many researchers have argued that the three foundations of reality consist of matter, energy, and consciousness, and that each of these components of reality enfolds the other two. They also contend that consciousness is an inherent part of reality, not just an abstract quality in the mind, and that reality has a holographic structure like that of the Mandelbrot set.

The universe, as well as all sorts of living and non-living entities, conforms to the fractal nature of reality. Any change to the universe anywhere affects other locations as well as the structure of the whole. In the double slit experiment, for example, the subatomic particles behave in strange ways and can manifest either as a wave or as a particle. Moreover, the act of measuring or observing the particles affects the outcome of the experiment, and leads to the collapse of the wave function. Here we emphasize the role which the observer or consciousness plays in affecting the behavior of subatomic particles. The similarities between the particles springing in and out of is without a doubt the result of observation and consciousness.

It is possible that this energy, which powers the entire universe, is consciousness projected through the interference pattern of energy waves and is

---

[87] Bieberich, E. 2008. "Structure in Human Consciousness: A Fractal Approach to the Topology of the Self Perceiving an Outer World in an Inner Space." Accessed August 06, 2017. http://cogprints.org/79/1/struc2.htm

responsible for what we perceive. A profound and thorough understanding of fractals helps us to unlock many paradoxes in our universe. Fractals appear to provide the foundation by which various levels of nature connect and separate, and acts as a bridge to link the inner and outer worlds.

What, then, is the origin of fractals, and why do they occur? The answer can possibly come from understanding the relationship between the continuous and discrete models. We know that atoms which make up everything we see around us are mostly empty inside. The subatomic particles within them are constantly interacting with one another, winking in and out of existence. The so-called solid perception of objects we perceive is merely the brain's interpretation of the electrical and biochemical signals it is constantly receiving via our senses.

What we think of as a solid object is merely an illusion, because our eyes and brains are designed to perceive complex structures as a single whole entity in order to help us understand what the world around us. This allows us to interact with our environment to make sense of the world in which we live and see things around us like houses, cars, food, etc. Our human eyes are designed to see continuous lines even though the objects we see are made up of atoms; we know that in an atom, the electrons that surround the nucleus move in orbits that occupy different levels and electrons can jump between discrete levels.

An electron that gains energy moves up a level while one that loses energy drops down a level. The energy is emitted as electromagnetic radiation of a certain frequency, this is the energy that surround us which some call the zero-point field, God, or consciousness. We learn from ST (as well as from QT) that particles that make up matter consist in turn of tiny of vibrating strings. As a result, if you change the vibrational frequency you create new particles. ST teaches us that subatomic particles are not point-like objects such as the one postulated in the SM, but rather are very tiny strings. These strings vibrate at different frequencies, and each vibration gives birth to a different particle.

Furthermore, advances in brain research and consciousness suggest that consciousness which inhabits every one of us is the source of all thought and, arguably, all physical matter. Consciousness as the origin of all thought is built from information transmitted by our mind. Many scientists now accept that body and mind are interrelated—despite our inability to explain the inter-relationship of body, mind, and consciousness in a logical and scientific manner. Advances in bio-psychology suggest that the mind is not the product of chemical reactions in the brain, but is an entity that thinks, perceives, discovers, and invents complex methods of interacting.

The mind does indeed function through the brain using nerve cells, nerve fibers, hormonal secretions, and electrical impulses. Our mind, with different patterns of thought, uses the brain and the body as the medium for its expression. It is possible that information creates fractals that form the basis of our HU. The physics of consciousness informs us that, once you amplify the frequency of your consciousness, the structure of the matter around you changes dramatically; this is the same as in ST and QT where if you change the vibrational frequency of mass you create new particles.

Consciousness is therefore not under the control of space-time. Energy-matter exists as a wholeness of discrete states; what we see around us is fractal states created by our mind, unlimited by space-time. Fractals are one of the most important keys to life and, therefore, reflect the multidimensional nature of consciousness. Is it possible to live in a universe without fractals? What would happen to matter around us without them? According to the laws of thermodynamics and entropy, the universe would possibly run down and we would witness chaos; entropy would increase dramatically and complete disorder would ensue. Particles would be scattered throughout the universe and interfere with our daily life. In fact, we each would not exist as a human being. Here we can only emphasize the importance of fractals. We would also like to stress that our perception manifests fractal organization.

For example, a tropical fruit such as a mango appears to us as a whole indivisible entity or object, however, when we cut it in half we can see that it is made up of different layers. It is well structured and contains, seeds, fluid, glucose, etc.; each of the tiny parts that make up the mango interact with one another. A smaller part of matter contains replicas, no matter how small, of the rest. These facts illustrate why we consider that fractals form the hidden structure of our universe. Perhaps the secrets of the universe are enfolded in fractals. One can do no better than quote Noel Huntley[88]:

> *It is now particularly relevant to introduce the fact that perception manifests fractal organization. For example, if one gazes at a clump of grass at certain distances away, the blades of grass will appear to group, mentally, more than they actually do physically. The attention is of course jumping around maybe erratically through different degrees of focuses but any groupings which exist in the environment, such as a whole house, causes the perception to focus on gestalt states. The intensity of attention is uniform across such a gestalt vision--it is one whole (frequency). All that is perceived is always in gestalt format whether it is small or large.*
>
> –HUNTLEY 2001

Fractals in Space-Time and Cosmology

I would now like to present an overview of the concept of fractals in cosmology, how it evolved, and highlight the theoretical and observational evidence that supports the idea that we live in a fractal universe. I apply the concepts of fractal geometry we have learned and draw some similarities between it and the universe by stressing the role of fractal geometry in an interdisciplinary mathematical approach to cosmology. A detailed account of the concept of a hierarchical cosmos and its development, as well as the idea of a fractal cosmos, with a review since ancient Greek times to the present, has been elaborated by Grugic[89].

Before looking at the idea of fractals in cosmology, I would like to say a word about the dominant cosmological model that was based on Einstein's

---

[88] Huntley, N. 2001. "The Source of Fractals." Accessed April 10, 2010. http://www.fractal.org/ Bewustzijns-Besturings-Model/Source-of-Fractals.htm

[89] Grujic, P. V. 2002. "The Concept of Fractal Cosmos: II. Modern Cosmology." *Serbian Astronomical Journal* 165 (2002):45–65 UDC 524.83

cosmological principle. Just after developing his theory of GR, Einstein decided to apply it to the cosmological problem—this was an attempt to build and understand a new model of the universe. He designed a model filled with stars and proposed that the stars have a natural uniform spatial distribution. He went on to argue for the principle of no center, and the uniformity of matter distribution and geometry of space, contrary to the arguments of those who proposed a hierarchical distribution of stars. In the early 1900s (before Einstein's proposed model), Karl Schwarzschild made similar arguments in his discussion of stellar parallaxes and the curvature of space. By 1922 Friedman had built on and expanded the work of both Einstein and Schwarzschild, and he went on to propose the theory of the expanding universe.

It is worth mentioning that, at the time, there was hardly any research being done on galaxies. As a result, there was no direct observational evidence for Einstein and Schwarzschild's proposed Cosmological Principle hypotheses. Their model was based only on theoretical reasoning. It did not take long before events took a new direction following the discovery of galaxies. Edwin Hubble[90] proposed a slightly modified version of the Cosmological Principle. Other scientists and cosmologists proposed different versions of the same principle. For instance, Milne came up with a modern version which states that "The whole world-picture as seen by one observer (attached to fundamental particle or galaxy) is similar to the world-picture seen by any other observer..." In a nutshell, the Cosmological Principle means that physical laws are the same everywhere and produce a similar outcome everywhere.

From the1930s onward observational evidence that would change the course of history started to accumulate. Edwin Francis Carpenter, then director of the Steward Observatory in Arizona, made a series of remarkable discoveries. He inspected clusters of galaxies and found out that their galaxy number density depends on the cluster size: the density is smaller in larger clusters. He discovered the density-size law: "That there is a universal

---

90 Hubble, E. 1929. "A Relation Between Distance and Radial Velocity among Extra-Galactic Nebulae." *Proceedings of the National Academy of Sciences* 15(3):168–173.

density radius law for the distribution of galaxies." Taking advantage of earlier work by Carpenter, Gérard de Vaucouleurs calculated the density of matter inside galaxy clusters of different sizes. He then proposed the existence of a universal density-size law, which is that

density = constant $\times$ sizess$^{-\alpha}$ where $\alpha \approx 1.7$.

In 1970 he summarized his ideas and connected the law of density with the concept of a hierarchical clustering, which formed the basis for a hierarchical cosmological model. Along with his wife Antoinette, they made immense contributions to the distances to galaxies and the value of the Hubble constant. They also found evidence for the flattened local supercluster around the Virgo cluster of galaxies. De Vaucouleurs's[91] ideas were received with little enthusiasm and rejected by those who continued to support the Cosmological Principle. It took several years, until the work of Benoit Mandelbrot on the fractal structure of the galaxy, for hierarchical cosmology to take shape.

As far back as the early 1900s, one of the proponents of the hierarchical universe, Fournier d'Albe, proposed an ingenious scheme of a hierarchically arranged cosmos which he divided into three levels. The zero level was made up of galaxies, the first level was made of a collection of spherical clusters, and the third level contained spherical superclusters. In fact, his original scheme was not much different from a multiverse plan. At about the same time, Felix Hausdorff made remarkable progress—well illustrated in the number of papers published—where he proposed a generalization of topological notions such as "metric, measure, dimension, etc."

Many years then passed without much progress. Later, with the development of the theory of fractals, mathematical objects that have a Hausdorff dimension, light was shed on many of the previous cosmological models, particularly the concept of the hierarchical universe. As we know, two parameters characterize fractal systems or objects: the fractal dimension D (also

---

[91] Vaucouleurs, G. de 1970. "The Case for a Hierarchical Cosmology." *Science* 167(3922):1203–1213.

known as Hausdorff's dimension) and the lacunarity. The latter determines the exact geometrical structure of the system within the class of objects that possess the same fractal dimension.

Astronomical evidence accumulated since the early 1930s supports the idea of a hierarchical cosmos. In 1934 Shapley published a survey that showed clusters of galaxies even larger than the Virgo cluster. The work of Fournier d'Albe showed the larger concentration of clusters and superclusters. However, it was more than 30 years before the immense contributions made in 1970 by de Vaucouleurs, who identified a pattern of galaxy clustering and made an estimation of their fractal dimension, changed the thinking of the majority of cosmologists.

The beginning of the end for the Cosmological Principle came in the form of a book by Mandelbrot[92], published in 1977, entitled *Fractals: Form, Chance, and Dimensions,* in which he showed that galaxies are fractally distributed. He introduced fractals into cosmology by generalizing Einstein's Cosmological Principle (corresponding to D=3), which allows a non-uniform galaxy distribution with D<3. Further progress was made in 1983 when Benoit Mandelbrot outlined his new ideas and decided to apply the concepts of the fractal to cosmology; he introduced the notion of lacunarity, which took into account the presence and role of cosmic voids which determine the topology of the fractal system beyond the fractal dimension D.

It is worth pointing out that the work of Mandelbrot, coupled with the introduction of the concept of fractals, that allowed hierarchical structuring to attain a more fundamental footing. It took exactly four years so that in June 1987 the Cosmological Principle, which had been dominant for years, came to an end. The work of Luciano Pietronero[93] contributed heavily to this shift with his now famous article entitled "The Fractal Structure of the Universe:

---

92 Mandelbrot. B. 1997. *Fractals: Form, Chance, and Dimensions*. W.H. Freeman & Co. Ltd.: New York.

93 Pietronero, L. 1987. "The Fractal Structure of the Universe: Correlations of Galaxies and Clusters and the Average Mass Density." *Physica A: Statistical Mechanics and its Applications* 144(2–3):257–284.

Correlations of Galaxies and Clusters and the Average Density." However, despite the work of Pietronero and others, it took more than 12 years of active research until, in January 1999, a paper in *Nature* acknowledged that the universe is essentially fractal on scales below 100 Mpc.

The fundamental importance of fractals in cosmology has been pointed out by one of his ardent defenders, Pietronero. One can do no better than quote from Grujic[94]:

> *From a more formal, mathematical point of view, since the concept of fractal has emerged in the second half of the last century; hierarchical structuring has attained a more fundamental ground. In particular, the advance of the topology of space-time has enabled us to make a clear distinction between the standard matter distributions, which are described by well-conceived analytical functions and fractal, nonanalytical ones. The concept of a Copernican principle has gained a new significance within the fractal pattern. Universe may be inhomogeneous, but still, it will appear the same (isotropic) to every observer.*
>
> —GRUJIC 2003, 262

Observational and theoretical evidence in astronomy has shown that the universe has a large-scale structure and that this structure might be fractal or hierarchical. In cosmology, observational evidence from a wide range of redshift surveys suggests that the fractal dimension of large-scale galaxy clustering is roughly $D_F \sim 2$, providing more support for current theories of holographic cosmologies. In addition, many physicists believe that the universe itself has a fractal shape, due to its foam like structure and the enormous globular voids between clusters and galaxies.

It is worth pointing that, despite the progress made, the fractal paradigm has not yet been recognized by most physicists, particularly cosmologists.

[94] Grujic, P. V. 2002. "The Concept of a Hierarchical Cosmos." *Publications of the Astronomical Observatory of Belgrade* 75 (2003):257–262.

However, recent observational and theoretical advances related to the global cosmic structure may lead to a new understanding of our universe. I am convinced that if the alleged fractal dimension turns out to be a whole number, i.e., D_F=2 this will hint at the presence of a holographic structure in our universe. The various independent estimations of the galaxy clustering dimension in Table 3.1 suggest that D_F has a value around 2.

The pioneering work of de Vaucouleurs, Einasto, Zwicky, Pietronero and many others continues to be a great source of knowledge and information, in particular their work on the observable galactic redshifts which have been measured using the Hubble law. Furthermore, the results of redshifts are well documented by the Center for Astrophysics Survey (CFA), the Southern Sky Redshift Survey (SSRS), the Las Campanas Redshift Survey (LCRS), and the European Southern Observatory Slice Project (ESP). For instance, the LCRS team have used multi-object spectrographs to measure redshifts of up to 120 galaxies simultaneously; their results in the form of 3D maps show the space distribution of galaxies.

These results are compelling and demonstrate that the case for a HU is strong. The success of fractal geometry continues to shed light on the structure of our universe and is perhaps leading us a step closer to understanding the process that creates large-scale cosmic formations.

Table 3.1 *A compilation of galaxy fractal dimension calculations for various redshift surveys. (Adapted from Mureika 2007, 2.)*

Survey	Fractal Dimension (D_F)	Total Number of Galaxies (approx. size)
CfA1	1.7±0.2	1800
CfA2	~2	11000
CfA	1.9±0.2	1845
SSRS1	2.0±0.2	1700
SSRS2	~2	3600
SSRS	2.0±0.2	1773
LEDA	2.1±0.2	75000

Survey	Fractal Dimension (D_F)	Total Number of Galaxies (approx. size)
LEDA	2.1±0.2	70000
LCRS	1.8±0.2	26000
Las Campanas(LCRS)	2.2±0.2	25000
Sloan	2.1±0.2	1000 000
IRAS 1.2/2 Jy	2.2±0.2	5000
Perseus-Pisces	~2.1	3300
ESP	1.8±0.2	3600
ESP	1.9±0.3	4000
SDSS(r1)	~2	$2\times10^5 - 1.5\times10^6$
2dF	2.1±0.2	250 000

PART 4

The Prospect of a Holographic Universe

Implications Beyond Theoretical Physics

Indeed, the attempt to live according to the notion that the fragments are really separate is, in essence, what has led to the growing series of extremely urgent crises that is confronting us today.

—DAVID BOHM

CONCLUSION

We have looked at the meaning of the HU from various viewpoints of science. It is now time to sum up by looking at its implications for theoretical physics and beyond. Despite the achievements and successes of QT, some deficiencies of the theory have been pointed out. These include the wave-particle duality, the double-slit experiment, entanglement, the Heisenberg Uncertainty Principle, the quantum measurement problem, non-local effects, etc. These unresolved issues and fundamental problems suggest that our universe is operating on many different levels rather than on a three- or four-dimensional reality. A consensus is emerging that QT explains the nature of the multidimensional reality of our universe.

The issue of quantum entanglement or quantum non-locality has been a recurring one. It has, in fact, inspired some physicists to come up with answers; perhaps one of the more radical explanations that we have put forward in this book is the HU. The late theoretical physicist David Bohm pointed out that we are all connected to one another and that there is no separation or distance between all that exists. It is more likely that at some deeper level of reality, a particle is not an individual entity, but merely an extension of the same fundamental entities. We are all linked and connected with everything else in the universe. Many other scientists from various fields such as Aspect, Pribham, Bekenstein, Gregus, George Dela Warr, Schempp, Smolin, etc. suggest that despite its solidity, the universe is at heart a gigantic hologram.

The discoveries of non-locality and wave-particle duality suggest that everything in the universe is joined or connected. Most basic atomic particles comprise the very fabric of the material universe. Space and time are composed of the same essence as matter, and these particles may also be conscious. Matter and events interact with each other, as do space and time, and operate under the law of non-locality. Bell's theorem informs us that once connected, objects affect one another forever no matter where they are. If two objects are connected a stream of energy will always connect them forever. Recent advances in QT, QH, holonomic theory, QG, self-organization theory, and cellular automata theory suggest that the entire universe is possibly holographically connected, an interlinked network of holograms, energy, and information.

The HU teaches us that during entanglement there is no energy transmission between particles or objects, but there is only information. As a result, it appears that there is no violation of Einstein's special theory of relativity. Entanglement simply means that we are all connected and that this wholeness is not subject to spatial separation. It is now clear that the hypothesis of a HU can be approached and understood in many ways in different fields of science and technology. One can see its manifestation and how the concept of a HU emerges. For instance, in plasma cosmology (electrons), QT (non-locality), QG (BH entropy), holonomic theory (brain), particle physics (lasers), fractal holography (fractal), QH (holography and non-locality), and biology (DNA).

In plasma physics, for example, Bohm showed (in experiments at Berkeley) that individual electrons self-organize and act as part of an interconnected whole. They behaved as though an original force was guiding them or they themselves were intelligent conscious entities. We have also seen in QG that the study of BHs suggests that our 3D reality is an illusion and is merely a holographic projection of what exists in two dimensions in the outer edges of the universe. The application of the concept of holograms came to be known as the HP, which combines the work and ideas of several physicists. The implication of this principle is that the potentially holographic nature of our universe derives from fundamental properties of the fabric of space and time.

We have learned from biology and brain research that this hypothesis extends to the human body as well. Each cell in a person has the complete DNA sequence required to produce an entire body—the whole is in the part and vice-versa. Every particle, living entity, or person contains all the information of the entire universe within itself. A comparison between the brain and a hologram demonstrates beyond any doubt the similarities and suggests that the brain processes information holographically.

On the problem of combining the two most successful physics theories, relativity and QT, scientists have struggled to come up with reasonable answers. Proposed solutions have varied depending on the school of thought. Despite the progress made so far, the failure to combine both theories suggests that there are flaws in our understanding of gravitational physics. It also hints at the possibility that the universe is digital or discrete and therefore, space and time are discrete too. The notion of continuous structures may have to be abandoned. The three different approaches to QG (BH thermodynamics, loop QG, and ST) suggest that space and time are discrete on the Planck scale.

Digital physics might lead us to the holy grail of physics, the ToE. It encompasses QG, QH, information theory, QI, quantum computation, self-organization theory, cellular automata, and some aspects of theoretical computer science. Although still in the making, one of its most important aspects involves extending QT. It is possible that when completed, it will provide answers to fundamental questions on the nature of reality, space, time, the origin of life, the meaning of life, and possibly, what is beyond our universe and others—or so we hope.

Over the last 70 years or so, our view of the universe has changed dramatically with the advent of QT, cosmology, DNA, consciousness, research on the human mind, computer science, the internet, information theory, quantum computation, QH, and quantum computing. In a nutshell, the role of information and computation, or the emergence of digital physics, has been emphasized. Our views (and the nature of reality) are being challenged by new findings that are leading us towards the belief that everything we see or experience around us is not a physical reality at all, but rather an interpretation of a multidimensional hologram, constructed and interpreted through our consciousness.

Recent experimental evidence (as well as past experiments) in various fields suggests that quantum non-locality is becoming more apparent in quantum systems as well as in living beings. Moreover, all the data and theoretical results in many fields of knowledge are pointing toward the idea that our universe itself is a holographic projection. That would be the ultimate proof that space-time, the fabric of the universe, is quantized or discrete.

The twenty-first century, thus, provides us with a unique opportunity (through the HU hypothesis) to solve some of the most pressing and challenging problems in science and society. If, because of our findings, our universe turns out to be a highly detailed hologram, the implications are profound. We can draw some important conclusions about the nature of reality:

- The information of the entire universe is contained in each of its parts. Every living entity, inanimate or animate, is ultimately connected to this hologram. The whole is an emergent phenomenon of the parts. The quality of the whole is determined by the nature of its components or constituents. The parts reflect the nature of the universe, but the nature of the universe is also a reflection of us.
- If everything is connected, separateness is an illusion: the past, present, and future are intimately linked. For instance, the electrons spinning in my body are connected to the electrons spinning in someone else thousands of miles away or on a star or planet located a billion light years away, etc. Thus, reality has a holographic structure.

- As a result, we ought to improve living conditions for all in society and end the discrimination committed against minorities, as well as the ongoing exploitation of the oppressed peoples of the world. This exploitation has been going on for the last 700 years or so through imperialism, colonialism, discrimination, neo- multicorporate colonialism, the stealing of local resources, minerals, lands, the destruction of local culture, languages, religion, gods, divinity, etc.
- We should work together to end the suffering, discrimination, and poverty that we see throughout the developing nations and in Europe, America, etc., where multinational companies have taken control of everything in the name of capitalism and are currently controlling the politico-social-economic sphere of life.
- The whole exists within each part. Consciousness is an important aspect of a living being that must be considered, and if our consciousness creates the physical world, then we must think carefully about every thought coming from our brains. Our actions and behaviors must be carefully measured. Human consciousness and the physical world cannot be taken as two different entities. Human thought, in fact, shapes our physical reality or the external world—we create the world around us and are therefore associated with our physical world as well as with others in the far cosmos.
- A change in consciousness is therefore necessary: a positive outlook, good thoughts, and actions, can change the physical world around us.
- The distance between things or objects is an illusion; solid objects that we see are just empty space because their mass is condensed energy, even though the world appears solid.
- All areas of knowledge and our way of living and interacting may have to change if these findings and experimental evidence turn out to be true. The implications are profound for religion, psychology, and other areas of art, science, and technology.

Consider the basic mathematical property of holographic systems in which the information of the whole system is distributed in each part of the scheme, as well as Pribham's brain holography system. In addition to Bohm's holographic quantum world, Bekenstein suggested that the universe could be like a gigantic hologram. Moreover, from the extension of Bekenstein's work by 't Hooft and Susskind, I propose that this universal interconnectedness shows that it is possible to access all the information (known or unknown) in the wave interference patterns that have existed in the entire universe since its beginning.

The quantum holoinformational nature of the universe interconnects each part of it—plants, animals, galaxies, planets, humans, brain-consciousness—with all the information stored in the holographic patterns distributed in the whole cosmos and other universes as shown in Figure 4.1.

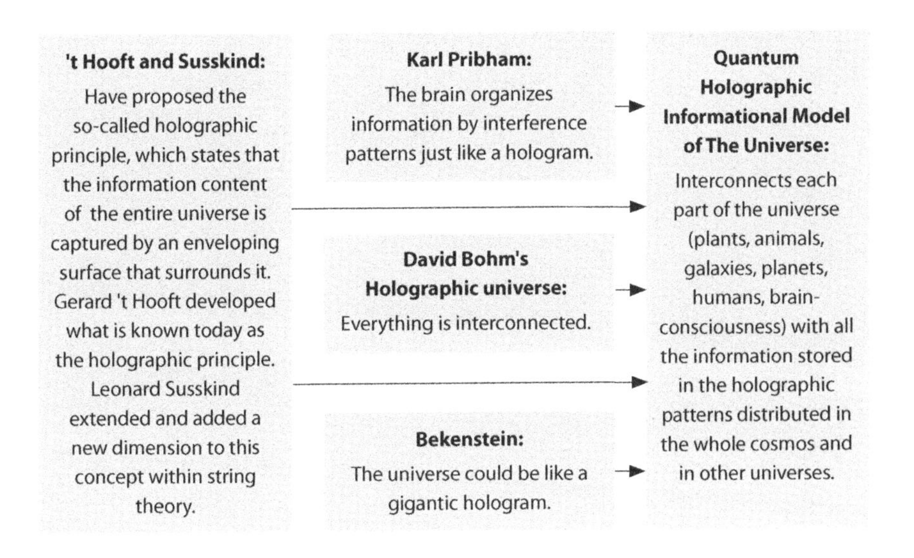

Figure 4.1 *The quantum holoinformational nature of the universe.*

GLOSSARY OF SCIENTIFIC TERMS

Aharonov-Bohm effect: Bohm and Aharonov showed how a magnetic field could influence the behavior of electrons confined far away from the field. Under certain conditions, electrons can feel the presence of a nearby magnetic field even if they are traveling in regions of space where the field strength is null.

Allais Effect: Anomalous behavior (movements) exhibited by a Foucault pendulum during the time of a solar eclipse.

Astronomy: Study of the universe as well as celestial objects, space, and time.

Atom: Fundamental unit or smallest particle of a chemical element. It is made up of a nucleus surrounded by electrons.

Background radiation: See cosmic microwave background radiation.

Bell's theorem: One of the ways put forward by J.S. Bell to test if QM was a local theory. Experiments indicate that QM is a non-local theory.

Big bang: Theory put forward that the universe originated as a singularity, i.e., from the explosion of dense matter.

Biophysics: Scientific studies of the application of the laws of physics to biological phenomena.

Black hole: Region of space with a gravitational field so intense that no matter, object, or radiation can escape it.

Black-hole entropy: The entropy embodied within a BH.

Brownian self-similarity: Is a technique used to make fractals more realistic. It is named after Robert Brown who discovered it while studying the motion of microscopic particles. See also self-similarity.

Cellular automata theory: CA theory suggests that everything in the universe is the result of interactions, the information changing state according to some specific mathematical rules. Cellular automata are discrete dynamic systems (mathematical discrete models). Each cellular automaton consists of an infinite regular array of cells (storage locations for numbers) or memory locations; each memory location contains some numbers. Those numerical data (1s and 0s for example) change state according to the rules of the cellular automata.

CERN: Conseil Européene pour la Recherche Nucléaire, French for European Organization for Nuclear Research.

Chaos theory: A branch of mathematics that studies complex or nonlinear systems, which are characterized by unpredictability and are impossible to control. Typical examples include turbulence, the stock market, the weather, etc. The idea behind the theory is to find the underlying order behind the apparently random data.

Cosmic Microwave Background Radiation (CMBR): Microwave radio emission composed of photons released from the primordial cosmic material.

Cosmology: The science concerned with discerning the origin, composition, and development of the universe. It combines astronomy, astrophysics, and particles physics as well as some analytical approaches mainly from geometry and topology.

Cosmological principle: Is the working assumption that, viewed on a sufficiently large scale, the properties of the universe are the same for all observers.

Consciousness: Refers to the state of being conscious, aware, or able to perceive. However, so far consciousness remains one of the least understood enigmatic properties of the human mind.

Dark energy: Is a possible form of energy that permeates all of space and tends to accelerate the expansion of the universe.

Dark matter (DM): Matter whose existence is inferred by dynamical studies. So far it has not been detected but it is predicted by many cosmological theories.

Dendron (dendrite): Refers to any of the threadlike extensions of the cytoplasm of a neuron. They compose most of the receptive surface of a neuron.

Deoxyribonucleic acid (DNA): Is the genetic material found in the nuclei of all cells.

Detector: Instrument or device for recording the presence of subatomic particles in an accelerator experiment.

Digital Physics: Is an alternative to statistical QM. Moreover, it is the study of discrete models for physics. It combines QM, quantum computation, QI, quantum holography, self-organization theory, thermodynamics, information theory, cellular automata, and some aspects of theoretical computer science. It is based on one of the following assumptions:

- The dynamic evolution of the universe is computable, i.e., the output of a computer program;
- The universe itself is a computer program run by an algorithm;
- The computer could be a huge cellular automaton **(Zuse, 1967)**, or a universal Turing machine **(Schmidhuber, 1997)**;
- Iinformation is more fundamental than matter and energy;
- Time and everything in space-time is discrete and can be modeled by discrete values like the integers.

Electromagnetic light: Light is a form of electromagnetic radiation released by various sources such as the Sun, flames, light bulbs, etc.

Electromagnetic waves: Consist of oscillating electric and magnetic fields which radiate outward at the speed of light. Typical examples include low-frequency radio waves (AM and FM), visible light, radar, infrared light, gamma rays, etc.

Entropy: A measure of the disorder of a physical system.

EPR experiment: A thought experiment designed to point out the inadequacies of QM. In 1935, a famous paper by Einstein, Podolsky, and Rosen, argued strongly about the strange behavior of entanglement and concluded that QM was an incomplete theory.

Euclidean Geometry: Mathematics that deals with regular shapes such as points, squares, lines, circles, triangles, etc. Euclidean space is flat, the three-dimensional analog of a plane. In contrast, non-Euclidean space is curved, such as the four-dimensional analog of a sphere or a hyperbola.

Explicate order: Is manifest. It includes the three-dimensional world of objects, space, and time. This explicate order is what can be perceived by our senses or by instruments; these are what is in the domain of physical reality. For instance, rocks, stars, the moon, planets, and ourselves.

Field: Environment in which the potential action of force can be described mathematically at each point in space.

Flyby Anomaly: Is an unexpected increase in acceleration experienced by some spacecraft during slingshot maneuvers. It has been observed that this unexpected energy increase during Earth flybys causes an unexplained velocity increase of over 13mm/s.

Fourier Transform (FT): Is a mathematical transform that expresses a mathematical function of time as a function of frequency. The result is known as its frequency spectrum.

Fractal: A rough or fragmented geometric shape that can be subdivided into parts, each of which is (at least approximately) a reduced-size copy of the whole.

Fractal geometry: Is a branch of geometry invented by Benoit Mandelbrot to study irregular and complex shapes that cannot be explored by Euclidean geometry.

Fractal Universe: A universe made up of fractals.

Gaussian probability: Is a continuous probability distribution that has a bell-shaped probability density function, known as the Gaussian function.

Galaxies: A large aggregation of stars, bound together gravitationally.

Geocentric model (Earth-centered): Old school of thought that placed the Earth at the center of the universe.

Geometry: Mathematics of points, lines, and shapes drawn through space.

Global cosmic structure: Refers to the fact that matter in the universe is not distributed randomly. This can be seen in the patterned distribution of galaxies, quasars, intergalactic gas, etc.

Gluons: Subatomic particles that are thought to bind quarks together.

Gravity: Interaction experienced by all mass or particles that possess mass.

Hierarchical Cosmology: A cosmological model in which the universe is arranged hierarchically.

Higgs boson: Particle believed to be responsible for giving mass to everything else.

Heliocentric model: Models that depicted the Sun as standing in the center of the universe.

Holistic: School of thought or philosophy that supports the view that natural systems and their properties should be viewed as wholes, not as collections of parts.

Holographic Universe: A theoretical model of reality that suggests that the physical universe is similar to a giant hologram.

Holometer: A sensitive laser interferometer designed to detect the smallest units in the universe called Planck units. Alternatively, its purpose is to detect the holographic noise.

Holonomic Theory: Proposes that the brain is similar to a hologram: in human brains the structure is holographic and the brain structures the senses holographically.

Hypothesis: A scientific proposition put forward to explain a given set of phenomena.

Implicate Order: Is the deeper, hidden, enfolded, the all-encompassing background to our daily experiences, physical, psychological, and spiritual.

Inflation: Theories that believe that the expansion of the early universe proceeded much more rapidly than it does today.

Kerr black hole: A spinning BH described by two physical properties: mass and angular momentum.

Landauer's principle: The erasure of information leads to an increase in the entropy of the universe.

Large Hadron Collider (LHC): Largest atom smasher located at CERN, where the physics of subatomic particles is studied. It contains a huge particle accelerator.

Laser: Light Amplification by Stimulated Emission of Radiation.

Liquid crystal: State of matter that has properties in between those of a conventional liquid and those of a solid crystal.

Locality: Assume that changes in physical systems require the presence of links between the cause and effect.

Logon: A logon or a quantum of information is the smallest area, in space and time, within which a signal can be encoded in the movement of energy and still maintain fidelity for information communication.

Mandelbrot set: A computer generated fractal created by Mandelbrot.

Microtubules: A component of the cytoskeleton, they are fibrous hollow rods that function primarily to help support and shape the cell. They are fundamentally important for maintaining cell structure and providing platforms for intracellular transport.

Molecule: The smallest unit of a chemical compound.

Non-locality: Refers to action at a distance or the direct influence of one object on another distant object.

Neuron: Also called a nerve cell. An electrically excitable cell that processes and transmits information by electrical and chemical signaling.

Neutrino: Electrically neutral particle.

Neutron: Neutral particle found in the nucleus of an atom – it consists of three quarks, two down quarks, and one up quark.

Occipital brain: Part or region in the back of the brain that processes visual information.

Phase Space: Refers to a space in which all possible states of a system are represented, with each possible state of the system corresponding to one unique point in the phase space.

Photon: Smallest bundle of light.

Pioneer Anomaly: Is a small sunward radial acceleration. In the particular case of Pioneer 10, the probe is moving as if it was subject to an unknown force pointing toward the Sun.

Planck energy: About 1,000 kilowatt hours, the energy necessary to probe to distances as small as the Planck length.

Planck length: Equal to roughly 10^{-33} centimeter, the size of a typical string in ST.

Plasma physics: A state of matter similar to gas in which a certain portion of the particles is ionized.

Quantum gravity (QG): A theory which attempts to develop scientific models that unify QM with general relativity.

Quantum Hologram: According to its proponents, it is a theoretical (i.e., imaginary) information-containing entity emitted by all physical objects (above the molecular level) which contains the entire history of the object.

Quantum information (QI): Physical information that is held in the state of a quantum system.

Quantum potential: In the Broglie–Bohm theory, the quantum potential is a term within the Schrödinger equation which acts to guide the movement of quantum particles.

Quantum theory: A theory that attempts to explain physical behaviors at atomic and subatomic levels.

Quarks: A fundamental particle that joins with others in triplets and doublets to form hadrons (protons and pi mesons).

Relativity: Encompasses two theories of Albert Einstein: special relativity and general relativity.

Self-organization: The evolution of a system into an organized form in the absence of external pressures.

Self-similarity: Exactly or approximately similar to a part of itself.

Shannon entropy: Is a measure of the quantity of information a system possesses.

Singularity: In cosmology, a point of infinite curvature of space where the equations of Einstein's general relativity break down.

Spherical Entropy Bound (SEB): The amount of information needed to fully specify the quantum state in a spherical region on its boundary, at a density on the order of one qubit per Planck area.

Standard Model: In the SM of particle physics particles are considered to be point-like objects.

String theory: Theory that replaces subatomic particles with strings. The strings are either closed loops or open. They vibrate in different ways, and the different modes of vibration give rise to all the different particles in the universe.

Subatomic particles: These are the particles which are smaller than an atom. There are two types of subatomic particle—elementary particles, which are not made of other particles, and composite particles.

Three-dimensional (3-D): A geometric (three parameter) model of the physical universe. It consists of three dimensions called length, width, and depth (or height).

Thermal equilibrium: Is a condition under which two substances in physical contact with each other exchange no heat energy. Thermal equilibrium is the result of the Zero-th Law of Thermodynamics: no heat flows between any two bodies which are at the same temperature.

Thermodynamics: Branch of physics that studies heat and other forms of energy in changing systems.

Thermodynamic entropy: Is a measure of the energy in a thermodynamic system which not available to do useful work.

Tully's objects: Objects discovered by astronomers that are older than the supposed age of the universe (Big Bang).

Two-Dimensional (2-D): It is a geometric model (two parameters) of the planar projection of the physical universe in which we live. The two dimensions are called length and width.

Wave-particle duality: Describes phenomena in quantum physics where quanta and other particles exhibit characteristics of both particles and waves.

Wilkinson Microwave Anisotropy Probe (WMAP): The aim of this mission is to determine the geometry, content, and evolution of the universe.

BIBLIOGRAPHY

Aharonov, Y. and Bohm, D. 1959. "Significance of Electromagnetic Potentials in the Quantum Theory." *Physical Review* 115:485–491.

Aiton, E. 1977. "Johannes Kepler and the 'Mysterium Cosmographicum.'" Accessed April 12, 2015. https://www.jstor.org/stable/20776469?seq=1#page_scan_tab_contents

Alvino. G. 1996. "The Human Energy Field in Relation to Science, Consciousness, and Health." Accessed April 12, 2017. http://www.vxm.com/21R.54.html

Amador, X. E. 2005. "Review on possible gravitational anomalies." *Journal of Physics: Conference Series* 24:247–252.

Anderson, J. D. Campbell, J.K. and Nieto, M.M. 2006. "The Energy Transfer Process in Planetary Flybys." Accessed November 08, 2011. http://arXiv:astro-ph/0608087

Anjamrooz, S.H. 2011. "Trinity is a Numerical Model of the Holographic Universe." *International Journal of the Physical Sciences* 6(2):175–181.

Aspect, A., Dalibard, J. and Roger, G. 1982. "Experimental Test of Bell's Inequalities using Time Varying Analyzers." *Physical Review Letters* 49:1804. Accessed May 01, 2013. http://www.phys.ens.fr/~dalibard/publications/Bell_test_1980.pdf

Beach, Fran k. A. 1961. *Karl Spencer Lashley 1890–1958 A Biographical Memoir.* National Academy of Sciences Washington D.C. Accessed September 18, 2014. http://www.nasonline.org/ publications/biographical-memoirs/memoir-pdfs/lashley-karl.pdf

Beck, T. E., and Janet, E. C. 2003. "A Quantum Bio-Mechanical Basis for Near-Death Life Reviews." *Journal of Near-Death Studies* 21(Spring):169–189.

Bekenstein, J. D. 2003. "Black Holes and Information Theory." *Contemporary Physics* 45:31–43.

Bekenstein, J. D. 2007. "Information in the Holographic Universe." Accessed February 02, 2016. https:// www.scientificamerican.com/article/information-in-the-holographic-univ/

Bekenstein, J. D. 1974. "Generalized Second Law of Thermodynamics in Black-Hole Physics." *Physical Review D* 9(12):3292–3300.

Bekenstein, J. D. 1974. "Black-Hole Thermodynamics." *Physics Today* 33:24–31.

Bekenstein, J. D. 2005. "How Does the Entropy/Information Bound Work?" *Foundations of Physics* 35(11):1805–1823.

Bekenstein, J. D. 1981. "A Universal Upper Bound on the Entropy to Energy Ratio for Bounded Systems." *Physical Review D* 23(2-15): 287–298

Bell, J. S. 1964. "On the Einstein Podolsky Rosen Paradox." *Physics* 1:195.

Benford, M. S. 2000. "Empirical Evidence Supporting Macro-Scale Quantum Holography in Non-Local Effects." Accessed September 6, 2011. http://www.Journaloftheoretics.com/Articles/2-5/benford.htm.

Benish, R. 2008. "Empirical Gap in Newtonian Gravity." Accessed September 15, 2010. http://www.gravitationlab.com

Bieberich, E. 2008. "Structure in Human Consciousness: A Fractal Approach to the Topology of the Self Perceiving an Outer World in an Inner Space." Accessed August 06, 2017. http://cogprints.org/79/1/struc2.htm

Bigatti, D. and Susskind, L. 2000. "TASI Lectures on the Holographic Principle." Accessed February 22, 2010. http:// arXiv: hep-ht/0002044v1

Bohm, D. 1981. *Wholeness and Implicate Order.* Routledge and Kegan Paul: London.

Bohm, D. and Hiley, B. 1993. *The Undivided Universe: An Ontological Interpretation of Quantum Theory.* Routledge: London.

Bohr, N. 1934. *Atomic Theory and the Description of Nature.* Cambridge University Press: Cambridge.

Bousso, R. 2002. "The Holographic Principle." *Reviews of Modern Physics* 74:825–874.

Bousso, R. 1999. "A Covariant Entropy Conjecture." Accessed October 12, 2010. http://arXiv:hep-th/9905177

Bousso, R. 1999. "Holography in General Space-times." Accessed October 11, 2010. http://arXiv:hep-th/0203101.

Buniy, R. V. and Hsu, S. D. H. 2007. "Entanglement Entropy, Black Holes and Holography." *Physics Letters B* 644(1):72–76.

Bradley, R. T. 2006. "The Psycho-Physiology of Entrepreneurial Intuition: A Quantum-Holographic Theory." In Proceedings of the Third AGSE International Entrepreneurship Research Exchange. February 8–10, 2006, Auckland, New Zealand.

Brunne, N., Branciard, C. and Gisin, N. 2008. "Can One See Entanglement"? Accessed July 07, 2010. https://arXiv.org/pdf/0902.2896.pdf

Brustein, R. and Hadad, M. 2009. "The Einstein Equations for Generalized Theories of Gravity and the Thermodynamic Relation $\delta Q = T\delta S$ are Equivalent." Accessed May 09, 2010. https://arXiv.org/abs/0903.0823

Broda, B. and Szanecki, M. 2009. "Induced Gravity and Gauge Interactions Revisited." Accessed July 13, 2010. http://arXiv:0809.4203v4 [hep-th]

Carroll, R. T. 2011. "Quantum Hologram." Accessed September 13, 2011. http://www. skepdic.com/quantumhologram.html

Chalmers, D. J. 1995a. "Facing up to the Problem of Consciousness." *Journal of Consciousness Studies* 2(3):200–19.

Chalmers, D. J. 1995. "The Puzzle of Conscious Experience." *Scientific American* 273(6):80–86.

Chalmers, D. 2010. *The Characters of Consciousness*. Oxford University Press: Oxford.

Cole, C. D. and Puthoff, H. E. 1993. "Extracting Energy and Heat from the Vacuum." *Physical Review A* 48(2):1562–1565.

Copernicus, N. 1543. *On the Revolutions of the Celestial Spheres* (edited by Jerzy Dobrzycki and translated and commentary by Edward Rosen). Polskiej Akademii Nauk: Wrocław, Poland

Czeslaw, H. 2010. "Non-local Compton Wave in Holographic Universe." *Prespacetime Journal* 1(6):992–996.

DeValois, K., DeValois, R. and Yund, W. 1979. "Responses of Striate Cortex Cells to Grating and Checkerboard Patterns." *Journal of Physiology* 291:483–505.

Di Biase, F. 2009. "A Holoinformational Model of Consciousness." *Quantum Biosystems* 3:207–220.

Di Biase, F. and Rocha , M. 1999. "Information Self-Organization and Consciousness." *World Futures—The Journal of General Evolution* 53:309–327, UNESCO, The Gordon and Breach Group, U.K.

Di Biase, F. and Rocha, M.S. 2000. "Information Self-Organization and Consciousness: Toward a Holoinformational Theory of Consciousness." In R. L. Amoroso et al. (eds.), *Science and the Primacy of Consciousness: Intimation of a 21st Century Revolution*, Noetic Press: Oakland.

Di Biase, F. 2009. "Quantum-Holographic Informational Consciousness." *NeuroQuantology* 7(4):657–664.

Easson, D. A., Frampton, P. H. and Smoot, G. F. 2010. "Entropic Accelerating Universe." Accessed June 05, 2011. http://arXiv:1002.4278v2

Elizalde, E. and Silva, P. J. 2008. "F(R) Gravity Equation of State." Accessed September 15, 2010. https://arXiv: 0804.3721v2 [hep-th]

Felten, J. E. 1984. "Milgrom's Revision of Newton's Laws - Dynamical and Cosmological Consequences." *Astrophysical Journal Part 1* (ISSN 0004-637X) 286:3–6.

Finzi, A. 1963. "On the Validity of Newton's Law at a Long Distance." *Monthly Notices of the Royal Astronomical Society* 127:21–30.

Fraser, B. 2008. "An Overview of the Nature of Time." Accessed June 02, 2010. http://www.fqxi.org/data/essay-contest-files/Fraser_NatureOfTime.pdf

Gao, S. 2010."Why Gravity is Fundamental." Accessed September 02, 2010. http://arXiv:1001. 3029v1 [physics.gen-ph][physics.hist-ph]

Gabor, D. 1946. "Theory of Communication." *Journal of the Institute of Electrical Engineers* 93:439–457.

Gabor, D. 1948. "A New Microscopic Principle." *Nature* 161:777–778.

Gariaev, P. P. and Junin A. M. 1989. "Phantom Leaf Effect. Myth or Reality?" *Energy* 10:46–52. In Russian.

Gariaev, P. P. 1994. *Wave based genome. Monograph.* Moscow. Ed. Obshestv. Pol'za. 279p. In Russian.

Gersl, J. 2008. "Bousso Entropy Bound for Ideal Gas of Massive Particles." *International Journal of Modern Physics D* (18):763–779.

Grujic, P. V. 2002. "The Concept of Fractal Cosmos: II. Modern Cosmology." *Serbian Astronomical Journal* 165 (2002):45–65. UDC 524.83

Grujic, P.V. 2002. "The Concept of a Hierarchical Cosmos." *Publications of the Astronomical Observatory of Belgrade* 75 (2003):257–262.

Haisch, B. 2001. "Brilliant Disguise: Light, Matter and the Zero-Point Field." Accessed February 05, 2015. http://altered-states.net/barry/update269/zeropointfield.htm

Hawking, S. W. 1976. "Black Holes and Thermodynamics." *Physical Review D* 13 (2):191–7.

Hawking, S. W. 1977. "The Quantum Mechanics of Black Holes." *Scientific American* 236:34–49.

Ho, M-W. 1999. "Coherent Energy, Liquid Crystallinity and Acupuncture." [Talk presented to the British Acupuncture Society]. Accessed January 12, 2012. http://ratical. org/coglobalize/MaeWanHo/acupunc.html

Ho, M-W. 1998a. "Organism and Psyche in a Participatory Universe." In D. Loye (ed.), *The Evolutionary Outrider.* The Impact of the Human Agent on Evolution: Essays in Honour of Ervin Laszlo (pp. 49–65). Praeger: Westport, CT.

Ho, M-W. 1998b. *The Rainbow and the Worm: The Physics of Organisms.* World Scientific Publishing: Singapore.

Hubble, E. 1929. "A Relation Between Distance and Radial Velocity among Extra-Galactic Nebulae." *Proceedings of the National Academy of Sciences* 15(3):168–173.

Jacobsongeneral relativity" , T. 1995. "Thermodynamics of Space-Time: The Einstein Equation of State." *Physical Review Letters* 75:1260.

Jeans, J. H. Sir. 1929. *Astronomy and Cosmogeny*, 2nd Ed. Cambridge University Press: Cambridge.

Galilei, G. 1632. *Dialogue Concerning Two New Sciences.* Accessed July 10, 2017. *http://galileoandeinstein.physics.virginia.edu/tns_draft/*

Greguss, P. 1973. "Holographic Concept in Nature." *In Holography in Medicine*. New York: Proceedings of the International Symposium on Holography in Biomedical Sciences.

Gryn, N. 2003. "David Bohm and Group Dialogue—or The Interconnectedness of Everything." *The Jewish Quarterly* Autumn: 93–97.

Gruji´c, P. V. and Pankovi´c, V. D. 2009. "On the Fractal Structure of the Universe." Accessed April 01, 2010. http:// arXiv: 0907.2127v1 [physics.gen-ph]

Holmes, C. P. 2011."The Origins and Nature of Human Consciousness, Part Three: Towards a Holographic Metaphysics of the Human Heart." *The Esoteric Quarterly*, Spring: 47–62.

Huntley, N. 2001. "The Source of Fractals." Accessed April 10, 2010. http://www.fractal.org/Bewustzijns-Besturings-Model/Source-of-Fractals.htm

Kak, S. 2008. "Indian Physics: Outline of Early History." Accessed September 02, 2010. http://arXiv:physics/0310001v1 [physics.hist-ph]

King, M. B. 1998. "Tapping the Zero-Point Energy." Accessed September 18, 2016. http://www.zamandayolculuk.com/tapzeropointenergy.htm

Lahanas, M. 2010. "The Myth of Newton's Apple, did Hipparchus Discover Newton Gravity and the Inverse Square Law?" Accessed June 02, 2010. http://www.mlahanas.de/ Greeks/ HipparchusGraviation.htm.

Lamoreaux, S. K. 2000. "Experimental Verifications of the Casimir Attractive Force Between Solid Bodies." *Comm. Mod. Phys. D: At. Mol. Phys.* 2:247–61.

Lashley, K. 1950. "In Search of the Engram." In *Physiological Mechanisms in Animal Behavior*. Academic Press: New York. pp. 454–482.

Laszlo, E. 1995. *The Interconnected Universe: Conceptual Foundations of Trans-Disciplinary Unified Theory*. World Scientific Publishing: Singapore.

Lee, J-W., Chan Kim, H. and Lee, J. 2010. "Gravity from Quantum Information." Accessed June 05, 2010. http://arXiv:1001.5445v2 [hep-th]

Maclay, G. J, Fearn, H. and Milonni, P. W. 2001. "Of Some Theoretical Significance: Implications of Casimir Effects." Accessed March 11, 2017. https://arxiv.org/pdf/quant-ph/0105002.pdf

Marcer, P. and Mitchell, E. 2000. "What is Consciousness?" *The Physical Nature of Consciousness,* P. Van Loocke (ed.), John Benjamins: Amsterdam. pp. 145–174.

Marcer, P. J. and Schempp, W. 1997. "Model of the Neuron Working by Quantum Holography." *Informatica* 21:519–34.

Marcer, P. J. and Schempp, W. 1998. "The Brain as a Conscious System." *International Journal of General Systems* 27:231–48.

Mandelbrot, B. 1977. *The Fractal Geometry of Nature.* W. H. Freeman and Company: New York.

Mandelbrot. B. 1997. *Fractals: Form, Chance, and Dimensions.* W. H. Freeman and Company: New York.

McFadden, C. and Skenderis, K. 2007. "The Holographic Universe." Accessed January 06, 2010. http:// arXiv: 1001.2007v1

Milgrom, M. 1998. "The Modified Dynamics—A Status Review." Accessed February 09, 2010. http:// arXiv: astro-ph/9810302v1

Mitchell, E. 2011. "What is the Quantum Hologram?" Accessed September 11, 2013. http://www.edmitchellapollo14.com/QHFAQs.htm

Mitchell, E. 1999. "Nature's Mind: The Quantum Hologram." Accessed September 11, 2013. http://www.nidsci.org/articles/naturesmind-qh.html.

Mitchell, E. and Robert, S. 2011."The Quantum Hologram and the Nature of Consciousness." Accessed January 02, 2016. http://journalofcosmology.com/ Consciousness149.html

Mitchell, W. C. 1997. "The Big Bang Theory under Fire." Accessed May 02, 2014. http://nowscape.com/big-ban2.htm

Moddel, G. 2009. "Assessment of Proposed Electromagnetic Quantum Vacuum Energy Extraction Methods." Accessed November 18, 2011. http:// arXiv: 0910.5893

Mureika, J. R. 2007. "Fractal Holography: a Geometric Re-Interpretation of Cosmological Large Scale Structure." Accessed April 10, 2010. http:// arXiv: gr-qc/0609001v2

Nishioka, T., Ryu, S. and Takayanagi, T. 2009. "Holographic Entanglement Entropy: An Overview." Accessed May 10, 2010. http://arXiv:0905.0932v2 [hep-th]

Ng, Y. Jack 2008. "Spacetime Foam: From Entropy and Holography to Infinite Statistics and Nonlocality." *Entropy* 10:441–461.

Nyambuya, G. G 2010. "Are Flyby Anomalies and the Pioneer Effect an ASTG Phenomenon?" Accessed August 18, 2011. http://arXiv:0803.1370

Padmanabhan, T. 2010. "Thermodynamical Aspects of Gravity: New Insights." Accessed April 12, 2015. http://arXiv:0911.5004 [gr-qc]

Padmanabhan, T. 2009." Equipartition of Energy in the Horizon Degrees of Freedom and the Emergence of Gravity." *Modern Physics Letters* A25:1129–1136.

Pais, A. 1982. *Subtle is the Lord.* ISBN-10: 0192851381, Oxford University Press: New York.

Penrose, R. 2007. *The Large, the Small and the Human Mind.* Cambridge University Press: Cambridge.

Pesci, A. 2008. "On the Statistical-Mechanical Meaning of the Bousso Bound." Accessed October 02, 2014. http://arXiv:0803.2642 [gr-qc]

Pesci, A. 2007. "From Unruh Temperature to Generalized Bousso Bound." *Classical Quantum Gravity* 24:6219–6226.

Piechocinska, B. 2004. "Wholeness as a Conceptual Foundation of Physical Theories." [Online] Accessed May 06, 2010. http: arXiv: physics/0409092v1 [physics.gen-ph]

Pietronero, L. 1987. "The Fractal Structure of the Universe: Correlations of Galaxies and Clusters and the Average Mass Density." *Physica A: Statistical Mechanics and its Applications* 144(2–3):257–284.

Pribham, K. H. 2004. "Consciousness Reassessed." *Mind and Matter* 2(1):7–35.

Pribham, K. H. 2006. "Holism vs. Wholism." *World Futures* 62:42–46.

Pribram, K. H. 1977. *Languages of the Brain.* Wadsworth Publishing: Monterey, CA.

Pribram, K. H. 1980. *Esprit Cerveau et Conscience.* In Science et Conscience, les Deux Lectures de l'univers. Editions Stock: Paris.

Pribham, K. H. 1991. *Brain and Perception: Holonomy and Structure in Figural Processing.* Lawrence Erlbaum Associates: Hillsdale.

Pribham, K. H. 1993. *Rethinking Neural Networks: Quantum Fields and Biological Data.* Lawrence Erlbaum Associates: Hillsdale.

Pribham, K. H. 1997a. "In Memoriam: Nobel Laureate Sir John Eccles." *The Noetic Journal* 1:2–5. Noetic Press: Orinda.

Puthoff, H. 1988. "Gravity as a Zero-Point-Fluctuation Force." *Physical Review* A 39(5):2333–2342.

Puthoff, H. 1994. "Inertia as a Zero-Point Lorentz Force." *Physical Review* A 49(2):678–694.

Puthoff, H. 1989. "Quantum Fluctuations of Empty Space: A New Rosetta Stone in Physics?" Accessed January 12, 2016. http://www.pia.com/Magazine/Ref/Puthoff_Rosetta_ZPE.html

Rees, M. 1997. *BEFORE the BEGINNING: Our universe and others.* ISBN: 0-684-81766-7, Simon & Schuster Ltd: U.K.

Rinpoche, S. 2008. *The Tibetan Book of Living and Dying: A Spiritual Classic from One of the Foremost Interpreters of Tibetan Buddhism to the West.* Random House: London

Rubin, V. C., Ford, W. K., Jr. and Thonnard, N. 1980. "Rotational Properties of 21 Sc Galaxies with a Large Range of Luminosities and Radii, from NCG 4605 (R = 4 kpc) to UGC 2885 (R = 122 kpc)." *Astrophysical Journal* 238:471–487.

Rueda, A. and Haisch, B. 2005. "Gravity and the Quantum Vacuum Inertia Hypothesis." *Annalen der Physik* 14(8):479–498.

Rueda, A. and Haisch, B. 1998. "Contribution to Inertial Mass by Reaction of the Vacuum to Accelerated Motion." *Foundations of Physics* 28(7):1057–1108.

Rueda, A. and Haisch, B. 1998. "Inertial Mass as Reaction of the Vacuum to Accelerated Motion." *Physics Letters A* 240(3):115–126.

Rueda, A. and Haisch, B. 1997. "Reply to Michel's Comment on Zero-Point Fluctuations and the Cosmological Constant." *Astrophysical Journal* 488:563–565.

Unruh, W.G. 1976. "Notes on Black Hole Evaporation." *Physical Review D* 14:870–892.

Sakharov, A. 1967. "Vacuum Fluctuations in Curved Space and the Theory of Gravitation." Translated from *Doklady Akademii Nauk SSSR* 177(1):70–71.

Schempp, W.1998. "Quantum Holography and Magnetic Resonance Tomography: An Ensemble Quantum Computing Approach." *Taiwanese Journal of Mathematics* 2(3):257–286.

Scientix. 2016. "DAVID BOHM, GENIUS OF MODERN SCIENCE IN SERCH OF TRUTH Explicate, Implicate Orders in Nature. Every Part of the Whole contains the Whole." Accessed February 18, 2016. http://tenets.zoroastrianism.com/DavidBohm GeniusofModernScienceinSearchofTruth.pdf

Simmons, G. S. 2007. *Calculus Gems: Brief Lives and Memorable Mathematics.* The Mathematical Association of America: USA.

Smolin, L. 2010. "Newtonian Gravity in Loop Quantum Gravity." Accessed March 07, 2010. http:// arXiv: 1001.3668v2

Strominger, A. and Thompson, D. 2004. "A Quantum Bousso Bound." Accessed May 04, 2010. http:// arXiv: hep-th/0303067 [hep-th].

Stonier, T. 1990. *Information and the Internal Structure of the Universe.* Springer-Verlag, New Addison-Wesley: Reading, MA.

Stuchebrukhov, A. A. 1996. "Tunneling Currents in Electron Transfer Reaction in Proteins. II. Calculation of Electronic Superexchange Matrix Element and Tunneling Currents Using Nonorthogonal Basis Sets." *Journal of Chemical Physics*107(16):6495–6498.

Susskind, L. 2003. "The Anthropic Landscape of String Theory." Accessed May 04, 2010. http://arXiv:hep-th/0302219v1

Susskind, L. 1995. "The World as a Hologram." *Journal of Mathematical Physics* 36:6377–96.

Svozil, K. 2008. "Microphysical Analogues of Flyby Anomalies." Accessed June 27, 2010. http://arXiv:0804.2198

Szeliskit, R. and Terzopoulos, D. 1989. "From Splines to Fractals." *Computer Graphics* (SIGGRAPH'89), 23:51–60.

Tabaro, J. P. 2011."African Heritage Series: Blacks Gave Science to Europe. "Accessed February 09, 2011. http://www.blackherbals.com/

Tabaro, J. P. 2008. "The Legacy of Ancient Egypt in Africa Today." Accessed February 9, 2011. http://onekamit.over-blog.com/tag/civilisations-verites et mensonges/.

Talbot, M. 1992. *Holographic Universe*. Harper Perennial: NY.

Theiler, J. 1990. "Estimating Fractal Dimension." *Journal of the Optical* Society *of America A* 7(6):*1055*–1073.

Thakur, S. 2010. "Quantum Entanglement and Holographic. "Accessed April 12, 2010. http://www.norlabs.org/

Thakur, S. 2010. "An Analysis of Big Bang Theory as a Work of Fiction." Accessed April 12, 2010. http://www.norlabs.org/

Thakur, S. 2010. "Human Eye can Instantaneously Detect Action at a Distance." Accessed April 12, 2010. http://www.norlabs.org/

't Hooft, G. 2000. "The Holographic Principle." Accessed September 02, 2010. http:// arXiv: hep-th/0003004v2]

't Hooft, G. 1993. "Dimensional Reduction in Quantum Gravity." Accessed October 11, 2012. http://arXiv:gr-qc/9310026.

Todd, J. 2010. "Disproving Induced Gravity and Induced Inertia Theories Related to Zero-Point Energy." Accessed June 16, 2016. http://www.gsjournal.net/Science-Journals/Research%20Papers-Relativity%20 Theory/Download/3910

Verlinde, E. P. 2010. "On the Origin of Gravity and the Laws of Newton. "Accessed July 07, 2011. http:// arXiv: 1001.0785[SPIRES].

Vaucouleurs, G. de 1970. "The Case for a Hierarchical Cosmology." *Science* 167(3922): 1203–1213.

Verlinde, E. P. 2000. "On the Holographic Principle in a Radiation-Dominated Universe." Accessed July 07, 2010. http://arXiv:hep-th/0008140v2

Vitruvius, M. P. 1914. "7". In Alfred A. Howard. *De Architectura libri decem* [Ten Books on Architecture, p. 215.]. VII. Herbert Langford Warren, Nelson Robinson (illus), Morris Hicky Morgan. Harvard University Press: Cambridge

Wikiquote. 2010. "Max Planck." Accessed July 07, 2010. http://en.wikiquote.org/wiki/ Max_Planck

Williams, C. 1996. "Metaphor, Parapsychology and Psi": An Examination of Metaphors Related to Paranormal Experience and Parapsychologicalr Rsearch." *Journal of the American Society for Psychical Research* 90:174–201.

Wright, A. E., Disney, M. J. and Thomson, R. C. 1990. "Universal Gravity: was Newton Right?" *Proceeding of the Acoustical Society of America* 8(4):334–338.

INDEX

F

G

H

I

J

L

M

P

Q

R

S

T

U

V

W

Z